Electrical Apparatus and Hazardous Areas

D1549794

Robin Garside

First Edition 1991
Reprinted with minor corrections and amendments 1992

ISBN 0 9516848 0 9

Published by Hexagon Technology Limited, Aylesbury, Buckinghamshire, UK. Printed in England by Unwin Brothers Ltd., The Gresham Press, Old Woking, Surrey.

Acknowledgements

A great number of people have, either directly or indirectly, given valuable assistance with this book. They are all included in this acknowledgement of thanks. I am especially indebted to the following for advice in their own areas of specialisation; Chris Towle and Dave Gaunt of MTL Instrument Group PLC, Maurice Hill of BASEEFA, Dave Wallden of Telektron Ltd.

Extracts from British Standards are reproduced with the permission of BSI. Complete copies of the Standards can be obtained by post from BSI Sales, Linford Wood, Milton Keynes, MK14 6LE; telex 825777 BSMIK G; telefax 0908 320856.

Finally, special thanks to Maggie for the hours of help at editorial stage.

Preface

For some time it has been clear that many people require a knowledge of the practices and procedures for electrical apparatus in potentially flammable atmospheres. Whilst there are many excellent technical publications, including National, International and European Standards available, these documents are often of detail interest only to the designer or person needing an in depth knowledge of a particular protection concept.

In this book I have attempted to fill the gap between the detail needed for a particular design and certification exercise, and the general information covered briefly in higher education courses, so as to provide a practical guide to the different methods of protection which are in common use throughout the world for electrical apparatus in hazardous areas.

Of course, any technical subject is continually evolving; new methods of protection emerge and old methods are refined and modified to take account of new and improved materials and techniques. Thus a reference book of this nature can never claim to be completely up to date. There are, no doubt, many people who will feel that publication of such a book should wait for the next edition of some particular Standard; but to do so would prevent publication at all! The reader should, however, ensure that he is familiar with the latest Standards and requirements of any particular application or installation with which he is involved, and should be aware that the requirements may vary from country to country. This book is intended to give general guidance and a base on which those needing specialised knowledge can build.

Many individuals and organisations have assisted me by providing guidance, examples of applications or help with layout, text and proof reading. My thanks, as always, to all of them. To the reader, I hope my effort will be considered worthwhile, and if there are omissions or suggestions for inclusions in future editions, please write and tell me.

Robin Garside
December 1990

Contents

Introduction

Safety, whatever the area of concern, is always a subjective matter. The precautions taken to achieve 'safety' will vary according to how the risk (of an accident or incident) is perceived, and according to the likely results of such an accident or incident.

Where human life is concerned, and where an incident could result in loss of life, the precautions which most people expect to be taken are quite stringent, especially if those at risk are not themselves wholly in control of the situation. Strangely, if they believe they are totally in control of a situation, they will in fact often take less safety precautions; a position which can, sadly, be observed daily by watching the way people drive their cars.

In industrial situations, the nature of the product or the processes involved often require human beings to work or live in close proximity to operations which are potentially extremely dangerous. This is an inevitable price we have to pay to enjoy the products of today's consumer world.

To reduce to zero the risk of an accident which would affect people would be totally impractical, since it would mean siting such plants at almost infinite distance from human habitation and, of course, operating the plant without any personnel at all. So we have to accept some risk, and the question quickly becomes: how much risk is acceptable?

For the purposes of this book, that question does not need to be answered directly, since there are established guidelines laid down in official publications; National, European and International Standards, Codes of practice and so on which, it must be assumed, indicate a level of risk (and thus a standard of safety) which is deemed reasonable.

It is also worth noting that reasonable safety is a moving target, and at present we live in a world where the 'acceptable risk' (at least once people are, for one reason or another made aware of the risk) is becoming smaller and smaller. That is to say people are, in their general perception of how other humans can affect them, prepared to accept less danger now than they would have done a few years ago. Perhaps the greater awareness and education over the past fifty years have a lot to do with this, and there is no doubt that in many things 'ignorance is bliss'. However, it is not just as simple as that, because even if safety precautions are available, they are not always used. I talked recently to a person who spent a good deal of his life in the mining industry, and he remembered that 'as a lad' no safety boots or helmets were available. He also remembered when they were introduced, and commented that the management had a hard struggle to persuade the miners to wear them! These days, safety boots and hard hats are provided just about anywhere there is any risk of danger, and we take such precautions for granted as perfectly reasonable.

Thus, the reader, who presumably has at least a passing interest in hazardous (potentially explosive) atmospheres and the possibility of electrical apparatus causing ignition of the hazard, will find that this book is firmly based on the advice and guidance of numerous Standards and Codes of practice. Because such documents are being continually revised and modified to incorporate improved safety requirements, the various methods of protection do alter as new editions of Standards and amendments are issued. It is most important to ensure that safety is reviewed in the light of the present thinking and generally accepted levels of risk. Thus there is no substitute for holding and being familiar with the current Standards and Codes as they apply to any particular installation and industry, and this book should not be regarded as an alternative to detailed knowledge. Rather, it attempts to explain the main methods of protection, and indicate in more practical terms than is possible in a published technical Standard, how these methods are used and the main safety criteria.

The reader should note that, unless specifically included by the text, the special requirements for hazards arising from the handling and manufacture of explosives and additional requirements for the mining industry are not covered by this book.

Section 1

This section covers aspects of general interest to all those involved with hazardous areas, and includes basic information on hazards, area classification, a short summary of each method of protection, and an explanation of Certification of apparatus.

CHAPTER 1

Hazardous Areas: Risks & Ignition Properties

Hazardous Areas: Risks & Ignition Properties

Hazardous areas are those places, commonly on industrial sites, where a potentially flammable atmosphere may exist. The flammable substance is usually a gas or vapour, but may also be a dust or fibre.

Before a fire or explosion of the flammable substance can occur, there must be sufficient oxygen to support combustion, and a source of ignition. The three constituents are often depicted in the form of an ignition triangle.

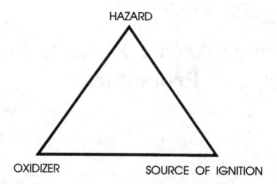

Figure 1.1 *The Ignition Triangle*

The **oxidizer** is normally air, and the limits of flammability and ignition data used throughout this book assume that air will be the supporter of combustion. Air contains about 20.95% oxygen. Thus any atmosphere containing more than about 21% oxygen, or where the partial pressure of oxygen exceeds 160 torr, must be considered as an **oxygen enriched atmosphere.** Since ignition can be more readily supported in such atmospheres, where they exist special precautions must be taken.

There is still a great deal of work to be done concerning oxygen enriched atmospheres, but as a general rule, wherever an oxygen enriched atmosphere may coexist with the presence of a potentially flammable substance, the factor of safety applied to any normal precautions should be at least doubled.

Table 1.1 shows in tabular form the relationship between height above or below sea level, and oxygen present in the atmosphere by % volume and partial pressure.

Total Absolute Pressure			Altitude Above or Depth below Sea Level	Partial Pressure of Oxygen if Atmosphere is Air	Concentration of Oxygen if partial Pressure of Oxygen is 160 mm Hg
Atmospheres	mm Hg	kPa	m Air or Sea Water	mm Hg	% by Volume
0.2	152	20	11735	32	100.00
0.33	253	33.8	8832	53	62.7
0.5	380	50.3	5486	80	42.8
0.66	506	67.6	3353	106	31.3
1	760	101.4	Sea Level	160	20.9
2	1520	202.7	10	320	10.5
3	2280	304.1	20	480	6.9
4	3040	405.4	30	640	5.2
5	3800	506.8	40	800	4.2

Table 1.1 [1] *Relationship Between Altitude and Atmospheric Oxygen Content*

As far as this book is concerned, the **source of ignition** with which we are concerned is, primarily, electrical apparatus. It should be remembered, however, that there are many potential sources of ignition of which the following may be representative:

Hot surfaces from process operations
Hot surfaces from heaters, including heating of buildings
Sparks from worn out safety boot steel
Sparks from scaffolding or steel ladders carelessly used
Naked flames, for example from people smoking
Magnification of sunlight from glass, for example a bottle
Frictional sparks from moving machinery
Static electricity sparks
Unauthorised welding operations
Lightning
Radio-frequency radiation[2]

In the main, such sources of ignition can and should be controlled by careful plant management, safety permits, operations procedures and constant attention to the layout, organisation and cleanliness of the plant.

Dealing with electrical apparatus is, however, not quite so easy. It is unreasonable to expect that an industrial plant can be run without the aid of electricity, since lighting, pumps, motors, instrumentation etc. will always be necessary. It should be remembered that whilst this book attempts to offer guidance on the various methods which may be used to give a reasonable level of safety from ignition for electrical apparatus, no method is foolproof. Where electrical apparatus exists, **there is always some risk that the electrical apparatus will become a source of ignition for any surrounding flammable atmosphere.** Thus, wherever possible, electrical apparatus should be removed from the hazardous area. This applies particularly when such apparatus could just as effectively be sited in an adjacent non-hazardous area, or where its purpose is no longer necessary and it can be decommissioned and removed altogether.

Hazardous areas are classified in terms of the type of hazard, and the likelihood of the hazard being present. The classification of hazardous areas is considered in more detail in Chapter 2. If an area is not a hazardous area, it is known as a **non-hazardous** or **safe** area. The terms are used interchangeably in this book, although the reader should be aware that some industries and authorities express a preference for one or other term. Since it is not uncommon to find notices at the entrance to hazardous areas stating 'use safe apparatus in this area' and suchlike, there would seem to be some possible confusion over the term 'safe area', and thus on balance the term non-hazardous area is probably preferable, since it is correctly self descriptive and less likely to be misinterpreted.

Hazardous Areas: Definitions and Terminology

The main definitions for hazardous areas concern the type of flammable substance which may be present: the **Gas Group** (IEC) or **Class** (North American NEC), the likelihood of the hazard being present: the **Zone** (IEC) or **Division** (North American NEC), and the **Temperature Class** or **Category.**

Gas Groups and Classes

The International Electrotechnical Commission (IEC) recognises two main Gas Groups:

Group I   Mining
Group II   Surface Industry (includes shipping and offshore installations)

Group II is subdivided into three sub groups, according to the minimum ignition energy of the hazard as follows:

Group IIC most easily ignitable: typically hydrogen
Group IIB less easily ignitable: typically ethylene
Group IIA less easily ignitable: typically propane

Thus hazards allocated to Gas Group IIC are the most severe hazards, Group IIB hazards less so, but more hazardous than those of Gas Group IIA.

Apparatus which is suitable for Gas Group IIC is also acceptable for Gas Groups IIB and IIA. Apparatus which is suitable for Gas Group IIB is also suitable for Gas Group IIA, but not for Gas Group IIC.

A comprehensive table of common compounds is included as Appendix 8.

Although North America (USA and Canada) recognise the IEC terminology and definitions, the system of hazard Classes is still in common use. There are three classes:

Class I	Gases and Vapours
Class II	Dusts
Class III	Fibres

Again, these classes are subdivided as follows:

Class I	Group A:	typically acetylene
	Group B:	typically hydrogen
	Group C:	typically ethylene
	Group D:	typically propane and methane

Class II	Group E
	Group F
	Group G

| Class III | No sub groups |

It is important not to confuse the two systems, since it will be seen that the sub group letters of the North American system and the IEC system are opposite in severity! Table 1.2 compares the two systems.

IEC SYSTEM	TYPICAL GAS	NORTH AMERICAN SYSTEM
GROUP I	METHANE	CLASS I GROUP D
GROUP IIA	PROPANE	
GROUP IIB	ETHYLENE	CLASS I GROUP C
GROUP IIC	HYDROGEN	CLASS I GROUP B
	ACETYLENE	CLASS I GROUP A

Table 1.2 *Comparison of IEC and North American Hazard Classification*

To add to the confusion, there are still frequent references made to obsolete classification systems, notably those used by British Standards BS 1259 and BS 229.

A comprehensive correlation of systems is shown in Table 1.3, with the IEC system, most commonly used, picked out in bold.

TYPICAL GAS	IEC 79	CENELEC	UK			GERMANY	USA
			BS 5345	BS 229 *	BS 1259 **	VDE0171	NEC 70 Class I Group
METHANE * * *	I	I	I	I	I	1	D
PROPANE	IIA	IIA	IIA	II	2c	1	D
ETHYLENE	IIB	IIB	IIB	IIIa or IIIb	2d	2	C
HYDROGEN	IIC	IIC	IIC	IV	2e	3a	B
ACETYLENE	IIC #	IIC #	IIC #	IV	2f	3c	A
CARBON DISULPHIDE	IIC # #	IIC # #	IIC # #	I V	2f	3b	-

NOTES

*   Obsolete Standard for flameproof enclosures

* *   Obsolete Standard on Intrinsic safety

* * *   Methane typifies mining hazards which are classified Group I. Where methane is a surface industry hazard, it is classed as Group IIA.

\#   Because of the very low limit of flammability, special precautions may be needed in an acetylene atmosphere.

\# \#   Because of the very low auto ignition temperature, special precautions may be needed for carbon disulphide.

Table 1.3   *Hazard Classification Systems*

Zones and Divisions

Once the nature of the potential hazard has been defined, the likelihood of the hazard existing needs to be considered. Clearly, there will be an infinite range of possibilities from 'definitely always present' to 'perhaps present for very short periods'. However, three distinctions are recognised. The term **zone** is used to describe the likelihood of the hazard existing at any given time.

Zones are defined by the IEC as follows:

Zone 0   Hazard continuously present or present for long periods.

Zone 1   Hazard likely to be present

Zone 2   Hazard unlikely to be present or only present for short periods of time, for example under fault conditions.

Many countries used to use the term **Division** rather than zone, and North America still uses this term. There are two Divisions:

Division 1   Ignitible concentrations of flammable gases or vapours exist under normal operating conditions, or may exist frequently because of repair or maintenance or because of leakage, or where breakdown or faulty, operation of equipment might release ignitible concentrations of flammable gases or vapours and might cause simultaneous failure of electric equipment.

Division 2   Volatile flammable liquids or gases are handled, but normally confined within closed containers or closed systems. Hazardous releases may occur under conditions of rupture of vessels or equipment. Or where concentrations are kept below flammable concentrations by mechanical ventilation, but may exceed such levels if the ventilation system fails. Or areas adjacent to a Division 1 area which may become hazardous by communication of the hazard from the Division 1 area.

Again, the IEC and North American terminology can be compared, although care should be taken since the term Division may have other implications when considering the selection of suitable apparatus. In general, unless there is good reason for doing so, for example on a North American plant, the IEC system should be used.

IEC ZONE	NORTH AMERICAN DIVISION
0	
1	1
2	2

Table 1.4 *Comparison of Zones and Divisions*

Temperature Classes

Whilst the Gas Group (or Class) relates to the ignition energy of the hazard, it is also necessary to have some way of classifying hazards for their ignition from hot surfaces. **There is no relationship between ignition energy and ignition temperature,** and hazards with a very low (dangerous) ignition energy, may have a relatively high ignition temperature.

(For example Hydrogen will, under worst case conditions, ignite with 20 micro Joules of ignition energy, and is placed in the most hazardous Gas Group: IIC. However, the ignition temperature of hydrogen is quite high at around 560°C.)

In order to establish the suitability of apparatus for use in a hazardous area from the view point of hot surfaces, apparatus is awarded a T-Rating, corresponding to its maximum surface temperature under certain conditions. The T-Rating or T-Class, can then easily be compared to the **Auto Ignition Temperature** or, **Spontaneous Ignition Temperature** of the hazard in which it is to be used, thus establishing safety from ignition from hot surfaces.

There are six temperature classes, with some sub classes recognised in North America. These are shown in Tables 1.5 and 1.6.

T-Class	Maximum Surface Temperature in °C
T1	450
T2	300
T3	200
T4	135
T5	100
T6	85

Table 1.5   *Temperature Classes (IEC)*

T-Class	Maximum Temperature in °C
T1	450
T2	300
T2A	280
T2B	260
T2C	230
T2D	215
T3	200
T3A	180
T3B	165
T3C	160
T4	135
T4A	120
T5	100
T6	85

Table 1.6   *Temperature Classes (North America)*

Again, there are some other systems still in use, although most modern apparatus and equipment will use the classification system shown in Table 1.5 above. However, for the sake of completeness, the German system which was commonly used until the introduction of the European CENELEC Standards is shown below, and compared to the IEC system.

IGNITION GROUP	LIMITING TEMP °C		TEMPERATURE CLASS	LIMITING TEMP °C
	CONTINUOUS	SHORT TIME		
G1	360	400	T1	450
G2	240	270	T2	300
G3	160	180	T3	200
G4	110	125	T4	135
G5	80	90	T5	100
-	-	-	T6	85

(VDE 0171 / IEC 79-0)

Table 1.7   *Comparison of VDE and IEC Temperature Classes*

Temperature classes make the assumption that the ambient temperature in which the apparatus is installed, will not exceed 40°C, unless stated otherwise on the apparatus label. Thus, the

T-Class can really be regarded as an indication of the temperature rise of the apparatus. (Figure 1.2)

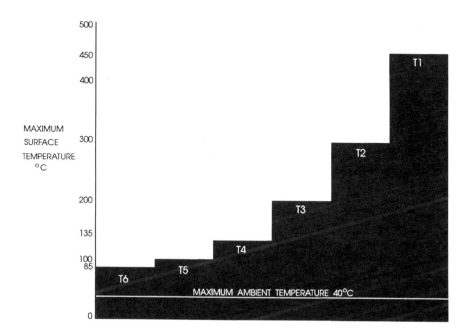

Figure 1.2 *T-Classes*

For those readers requiring a more complete discussion on the terminology and definitions of hazardous areas, Chapter 2 should be studied in depth.

14

Notes and References

Information from NFPA 53M, Fire Hazards in Oxygen
Enriched Atmospheres. 1979. National Fire Protection
Association. USA.

2 Detailed consideration of radio-frequency radiation ignition is
beyond the scope of this book. British Standard BS6656:1986,
entitled 'Prevention of inadvertent ignition of flammable
atmospheres by radio-frequency radiation', gives details of
preventative measures and measurement/calculation techniques.
Although RF certainly can give rise to ignition of flammable
atmospheres, the risk is generally small compared with other
sources of ignition.

CHAPTER 2

Area Classification

Area Classification

The Intention and Aim

The aim of area classification is to reduce, to an acceptable level, the possibility of a flammable atmosphere and an electrical source of ignition coinciding. For this to be achieved, the following information regarding the hazardous area must be available:

a) The nature of the hazard:

THE GAS GROUP

either IIA, IIB, or IIC, depending on the ignition energy of the hazard. The lower the ignition energy, the alphabetically greater the gas group.

For example:

Hydrogen (very easily ignited) = Gas Group IIC
Propane   (less easily ignited) = Gas Group IIA

Information on the Gas Group is usually determined by reference to tabular data published in, for example, the Code of practice.

b) The ignition temperature of the hazard:

THE T-CLASS

Hazards are classified according to their **auto ignition temperature A.I.T.** (sometimes called **spontaneous ignition temperature S.I.T.**) This is the temperature at which a substance will ignite without any external source of ignition, for example a spark or flame, being present. This data also is normally established from suitable tables, either in the Code of practice, or some reputable chemical handbook.

It should be noted that because a gas falls into a severe gas grouping like IIC, it does not necessarily follow that it will have a low A.I.T. For example Hydrogen has quite a high A.I.T., about 560°C.

A.I.T. should not be confused with **flash point (FP)**. F.P. is the temperature at which sufficient vapour is given off for ignition to take place **if a source of ignition is provided.** A.I.T., on the other hand, is the temperature at which the substance will self ignite. (That is without any extraneous source of ignition.)

c) The probability of the hazard existing:

THE ZONE

The zone defines how likely it is that a hazardous concentration will be present in any given geographical location.

There are three zones, 0, 1, and 2. (Zones used to be called Divisions, but this term is no longer used except in North America.)

The zone number stems from the source of release. If a source is continuously releasing, then this will lead to zone 0. If the release is likely to occur, then the zone will be zone 1. If the release is unlikely to occur, or will occur only for short periods of time, for example under some fault condition, then the zone number will be 2.

Sometimes, numerical values are associated with zones. These are not quoted in the Code of practice, but are well founded, and were used by ICI in their excellent publication 'Electrical Installations in Flammable Atmospheres' which, although published by RoSPA as of general interest, is sadly no longer available. The values however serve as a useful guide and are worth noting:

    Hazard present for   >1000 hrs / year = zone 0
    Hazard present for 10 - 1000 hrs / year = zone 1
    Hazard present for   <10 hrs / year = zone 2

These values approximate to:

Zone 0: 10% of the time or more
Zone 2: less than 0.1% of the time

The mathematics ties in with the risk of ignition from the different electrical methods of protection.

In addition to defining the zone number, the **extent of zone** needs to be established. This is the area over which the hazard will extend in concentrations such as to be ignition capable, **assuming that the source of release is actually releasing.** Normally, the extent of zone is the distance from the source of release to the locus of the **Lower Explosive Limit L.E.L.** (also sometimes known as the **Lower Flammable Limit L.F.L.**), although it may be taken as some nearby physical boundary - for example, the classification of an inside location may use the boundary walls as the extent of zone.

Thus area classification seeks to establish the nature of the hazard, Gas Group and T-Classification; and then to define the area over which the hazard may extend, the zone numbers and extent of those zones. In carrying out the study any opportunity of reducing or excluding hazardous releases should be taken.

The Effects of Area Classification

Although it will be clear from the above that the area classification study is not directly an electrical or instrument engineering function, the results of the classification are significant to the selection and use of electrical apparatus within the area.

The classification study may be used to control other aspects of hazardous area operation, for example the categories of personnel who may work in the area, but, according to the Code of practice BS 5345, and the IEC Standard 79-10, the area classification study is only associated with the electrical aspects of the plant.

There are two main criteria: the selection of electrical apparatus according to its method of protection, and the installation and maintenance of such apparatus.

Selection of Electrical Apparatus According to Zone of Risk

The Code of practice for the selection, installation and maintenance of electrical apparatus in flammable atmospheres, BS 5345: Part 1, lists the methods of protection which are suitable for each of the three hazardous zones. (See Pages 58-59).

It should be realised that since the Code is not a European document, the recommended methods of protection for the different zones are only applicable within the UK. Most other European countries follow the same general guidance, but there are some notable differences, for example oil filling and sand filling are acceptable for both zones 1 and 2 in European countries apart from Belgium and the UK.

Since the method of protection defines a statistical probability of ignition, and the zone defines the probability of the hazard being present, the use of, for example, a method of protection suitable for a zone 0 area in a zone 2, will lead to a greater level of safety in zone 2 than that applied in the zone 0. Unless there is some economic or other reason for using methods of protection which exceed the advice of the Code, there is probably little point in exceeding the recommendations, and the economic advantages which may be obtained from, for example using type N apparatus in zone 2, might as well be used.[1]

The Area Classification Exercise

The first stage in the area classification exercise is to examine all actual and potential sources of release of hazard and evaluate which, if any, can be reduced or removed altogether. Thereafter, the remaining sources of release need to be considered to define the zone number for the release. This is expressed pictorially in Figure 2.1.

20

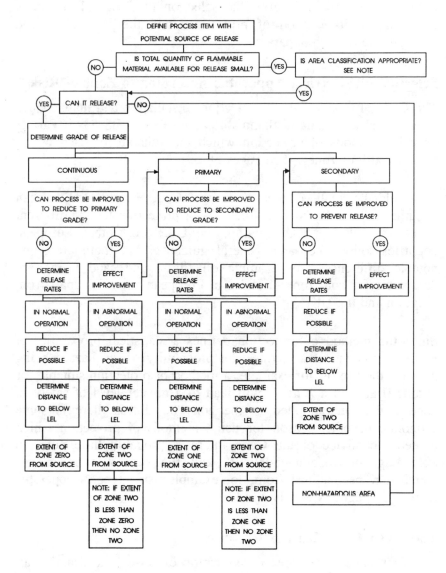

Note: If the quantity available for release is small, area classification techniques may not be appropriate, but other precautions may be necessary.

Figure 2.1 *Family Tree of Area Classification*

In order to keep the risk of ignition of any hazardous release within acceptable levels, the more severe the zone number, the greater is the protection required for electrical apparatus within the zone.

Before exploring the factors affecting the extent of zone in more detail, it is worth considering some other aspects of potentially flammable compounds. The following table [2] shows some properties of the hazard Hydrogen Sulphide, H_2S.

PROPERTY	UNITS	VALUE
Smell Detection Level	ppm	0.1
Safe Working Level	ppm	10
Serious Health Danger Level	ppm	50
Lower Explosive Limit	ppm	40000
Upper Explosive Limit	ppm	460000

It can be seen that, in this instance, a health hazard exists long before there is any danger of ignition. Whilst a similar situation exists for many substances, and although the human nose is often a very good gas detector, it cannot be relied upon since, as would be the case with Hydrogen Sulphide, high concentrations 'poison' the smell sense. Also, some hazards will not be noticeably sensed by smell until concentrations above the LEL are present. However, the table does demonstrate that very often the ignition danger is less significant than other hazardous considerations.

Consideration of Sources of Release: The Zone Number

The first stage in establishing the zoning information, is to establish the zone numbers for each source of release.

The normal approach is to examine each source of release or potential source of release in turn. Sometimes, a source of release under normal conditions may also present a larger source of release under abnormal conditions. Such situations lead to a relatively small zone 0 or 1, with a larger surrounding zone 2 area.

22

Taking that example, and supposing that it had been determined that under normal operating conditions, the hazard was likely to be released for say 500 hours per year, the resulting zone would be zone 1. Suppose that the extent of this zone had been established as a small spherical shape around the source.

The same source now needs to be considered under abnormal or fault conditions. **(But not catastrophic failure. Plant design should minimise the likelihood of catastrophic failure which could lead to a massive release. Catastrophic failure is not reconsidered at area classification stage.)** If, under these conditions, the release is greater, then the resultant zone size, now a zone 2 consideration, will be greater. With a knowledge of the properties of the hazard and the plant conditions, the extent of the zone 2 can be established.

Such situations give rise to a fairly common area classification pattern, with, for each source of release, a surrounding zone 1, with a larger zone 2. (Figure 2.2)

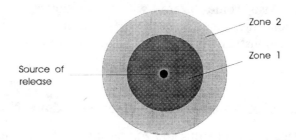

Figure 2.2 *Typical Assessment of Single Source of Release*

Clearly, on a complex plant, such a task may become impractical, and give rise to overlapping zones. For this reason, it is common practice to review the zones one by one, but then to take the resulting area as a whole. (Figures 2.3 and 2.4)

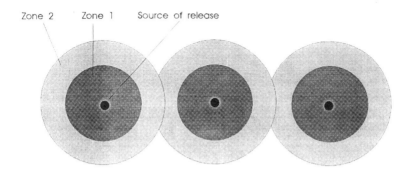

Zone 2   Zone 1   Source of release

Figure 2.3   *Initial Consideration of Several Sources of Release*

By considering several sources of release, we can see that the zone 2 areas overlap. The 'addition' of the zones does not significantly change the risk of the hazard being present, and a more realistic approach is seen in Figure 2.4.

Figure 2.4   *Result of Several Sources of Release*

Although this has actually increased the total hazardous area marginally, by smoothing out the curve, in practice this is most unlikely to cause difficulty.

A similar situation may occur inside a room where an open process vessel exists, for example a solvent tank. (Figure 2.5)

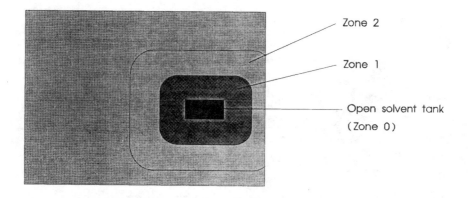

Figure 2.5 *Source of Release Inside a Room*

In this situation, it may be totally impractical to pronounce that the hazardous area is confined to within a defined line, and it would be quite understandable, especially if the zoned hazardous area represented a significant proportion of the room, to regard the entire room as a hazardous area, zone 2, with a zone 1 area local to the process vessel.

These two practical situations are commonly experienced, and common sense supports the position as explained. However, it does not in any way detract from the plant operators responsibility to **confine the hazard as much as possible.** This is probably the fundamental difference between modern area classification, and the older 'Division 1/Division 2' approach which tended to blanket classify large areas without a thorough analysis of each source of release.

It should be appreciated that the emphasis of current thinking on area classification is to eliminate zone 1 and zone 0 areas completely, or, where this is not possible, to limit the extent of such zones.

Consideration of Sources of Release: the Extent of Zone

Although, as discussed above, it may be convenient to take the extent of zone as some physical boundary, the true extent of zone should be established if possible, even if the result is not directly applied.

In order to establish the extent of the zone, the distance from the source of release to the lower explosive limit needs to be established. This task is virtually impossible to achieve with any degree of accuracy, since the result will be affected by many properties which are outside our control, and by properties which we know will vary. The distance to the LEL will, for example, be affected by ambient temperature, wind speed and wind direction, process temperatures and so on. It is therefore inevitable that there is a high degree of variability on the results of area classification studies on different plants of similar situations. Some specific industries have good additional guidelines, for example the Institute of Petroleum Code, to assist in the study, but many organisations are offered little help other than that available from the Code of Practice BS 5345, or the IEC document IEC 79-10.

Although there are a number of equations which may assist with the area classification task, because of the variables referred to above it is not reasonable to limit the study to an application of this mathematical approach. The classification study needs to be a mixture of previous experience, common sense, (both adjusted for specific plant conditions) and some mathematical work to give a confidence guide to the result. For ease of demonstrating this, a few examples may be useful. First however, a brief explanation of the calculations which may be involved.

There are often three stages of computation for consideration: defining the rate of release of the hazard, establishing the vapourization rate, and thus the distance to dispersion below the LEL. (Figure 2.6)

26

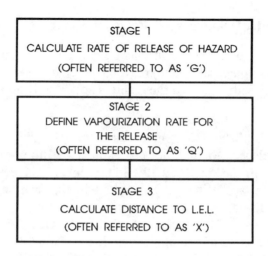

Figure 2.6 *The Three Stages for Establishing the Extent of Zones*

In some situations it is possible to go to the result more directly, but the same general principle applies.

Remember that although a theoretical distance to the LEL can be calculated for most situations, it is often possible to get a good guide by referring to published data and examples from documents such as the ICI/RoSPA Code, Institute of Petroleum Code etc., and if possible this approach should be pursued in preference to the use of numerous and complex calculations.

Frequently, a result calculated from dispersion equations, vapourization rates etc. will give an extent of zone which is smaller than that established by other means. This serves to emphasise that there is a tendency to over classify hazardous areas; that is to set larger zones than are actually required. However, unless there is a significant difference between the results obtained using different methods, (in which case a reason for the discrepancy should be investigated) the larger zone result will probably not present any practical difficulty and should normally be adopted.

There are numerous equations which can be used to calculate dispersions of gases etc. and, although the equations are based on sound chemical principles, the variability of behaviour of gases and vapours in certain circumstances may mean that alternative calculations will be more appropriate. The advice of a chemical engineer who is conversant with the behaviour of gases and vapours is essential if any significant degree of confidence is to be placed on calculated results. The equations[3] used in the following examples were first published, as far as the author can ascertain, in a proposed draft for the Code of practice BS 5345:Part 2, although they do not appear in the published Standard.

To demonstrate the use of the equations, and the interpretation of results, the following examples may be considered.

Example 1

Situation to be considered: pipe flange of process with methane piped under a pressure of 3.7×10^5 Nm^{-2}. (Figure 2.7.)

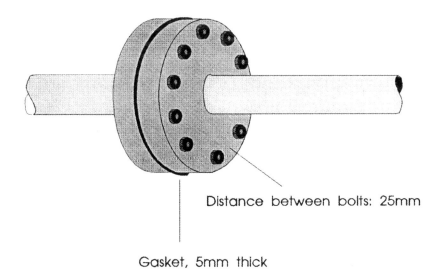

Distance between bolts: 25mm

Gasket, 5mm thick

Figure 2.7 *Pipe and Flange*

Data for Methane taken as:

Molecular weight   16
Process Temperature   298K
LEL   5% by volume

Under normal conditions there will not be any appreciable leakage. Thus there is no zone 0 or zone 1 area surrounding the gasketed flange.

However, it is necessary to consider abnormal operation, and this could be a situation where a section of the gasket between two supporting bolts weakens, breaks and blows out.

The resulting cross section leakage area will be

$$0.005 \times 0.025 \text{ m}^2$$
$$= 0.000125 \text{ m}^2 \qquad (1.25 \times 10^{-4} \text{ m}^2)$$

Referring to Figure 2.6, in this case the appropriate equation for 'G':

G $= 0.00586 \times A \times P_1 \times (M/T_1)^{0.5}$ Units: kg s^{-1}

Where:

A   = Cross section of leak path in m^2
P_1 = Absolute pressure upstream of release in Nm^{-2}
M   = Molecular (formula) weight
T_1 = Absolute temperature upstream of release in K

Valid for $P_1 > 200000$ Nm^{-2}

gives a result for of 0.0628 kg.s^{-1}, which, substituted in an equation to determine 'Q':

Q $= 0.0821 \left(\dfrac{G \times T_1}{M} \right)$ Units: m^{3}s^{-1}

Where:

G   = Mass flow rate from above
T_1 = Ambient temperature of released material in K
M   = Molecular weight of vapour

gives Q   = 0.09603 $m^3.s^{-1}$. This value, applied to an equation for 'X':

Release from High Level Point Source

$$X = \left(\frac{920 \times Q}{E}\right)^{0.55} \quad \text{Units: m}$$

Where:

Q   = Volume flow rate in $m^3.s^{-1}$
E   = Lower explosive limit in % Vol

gives the distance to the LEL, X = 4.86m from the source.

The ICI/RoSPA code gives a similar example, and shows a diagram as reproduced in Figure 2.8.

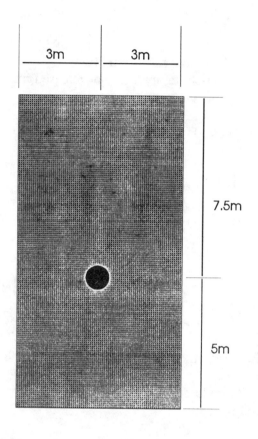

Figure 2.8   *Leakage from a Pipe Flange Hazard Lighter than Air*

Clearly, the ICI/RoSPA code has made more allowance for the fact that the hazard is lighter than air since it shows the upward vertical extent of zone as greater than the downward vertical extent of zone. Also, the code does not specify the size of leak or the flow rate, so is intended to be representative of a typical situation.

In order to understand the effect of leak size on the result, consider the graph of Figure 2.9.

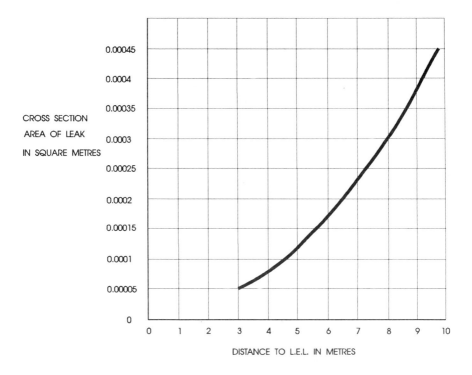

Figure 2.9   *Theoretical Effect of Leakage Hole Size on Example 1.*

This seems to clarify matters somewhat, and goes some way to justifying the case given in the ICI/RoSPA code.

Similar documents can be used to find examples which are similar to situations encountered on most plants. A list of useful codes and documents is given at the end of this chapter.

Example 2

Now consider a spillage of Ethanol (Industrial Methylated Spirits) giving a pool of the hazard of area 10m^2, with a diameter of 1m. Assume the temperature of the ethanol pool is 25°C.

32

In this case, it is possible to directly apply an equation to determine 'X':

Release From Pool of Liquid Below Atmospheric Boiling Point

$$X = \left(\frac{0.698 \times 10^9 \times A \times T_1 \times P_{V(T1)}}{d}\right)^{1.14} \times \left(\left(\frac{100}{E}\right)^{1.14} - \left(\frac{1}{P_{V(T1)} \times 10^5}\right)^{1.14}\right)$$

Units: m

Where:
A = Surface area of pool in m²
$P_{V(T1)}$ = Vapour pressure at T_1 in Nm⁻²
d = Diameter of the pool perpendicular to wind direction in m
E = Lower explosive limit in % Vol
T_1 = Temperature of pool in K

and using a temperature of 298K, a vapour density of 1.59, vapour pressure of 7968 Pa at 298K, and LEL of 3.3% by volume, a distance of 0.3m to the LEL is calculated.

The closest situation given in the ICI/RoSPA code is for a separator sump. (Figure 2.10) The pictorial solution shown needs to be modified for the situation considered, and so the zone 0 and 1 can be disregarded.

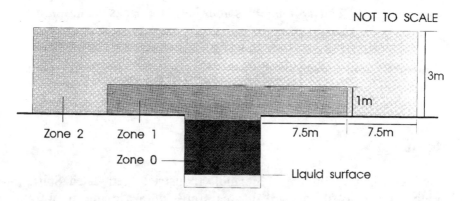

Figure 2.10 *Classification of Sump area*

This time, an allowance has been made for the hazard being heavier than air. (Vapour Density of ethanol = 1.59, compared to air = 1)

Even so, at first sight the code seems somewhat pessimistic, but consideration of the effects of temperature on the calculation give a truer picture as indicated in Figure 2.11.

It must be remembered that the position may be worse for some chemicals (e.g. acetone). The Code does, however, include a note which states that the figure is a general case and smaller distances may apply to situations where the hazardous material can be accurately defined.

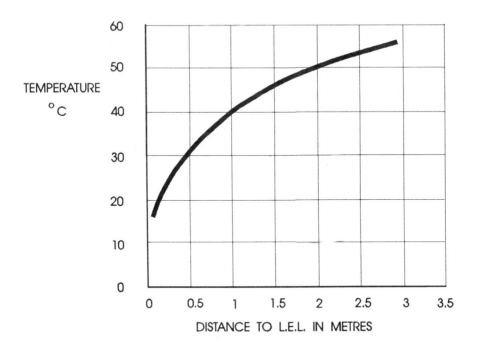

Figure 2.11   *Consideration of Example 2 (Ethanol) With Respect to Temperature*

Ventilation and its Effect on Zones

Ventilation, both natural and forced, is often considered important when designing industrial sites. The effect of ventilation, especially forced ventilation, is often misunderstood with regard to area classification.

Since the definition of the zone number relates to the type of release from which the hazard will stem (continuous release = zone 0; primary release = zone 1; secondary release = zone 2) it should be clear that no amount of ventilation can change the **zone number**. [4]

Ventilation may be used to locally extract fumes from a local source, thus reducing the extent of zone around the source, or, by increasing the number of air changes per unit time within an area it may be possible to show that the distance from a source of release to the LEL is reduced. Thus, in certain circumstances ventilation may alter the **extent of zone.**

It should be remembered that the concentration of vapour within a ventilation system may itself be such that the ventilation ducts are regarded as hazardous zones, and thus any sensors, motors, fans etc. within the duct must be suitable for that zone application.

Also remember that the outlet point of any duct may be a source of release. Unless it can be shown that the air flow is diluting the hazard to below the LEL, then the ventilation may just be transferring the problem to somewhere else!

Depending on the nature of the release in terms of the source and the volume of hazard, it may be necessary to link the ventilation system into some alarm system, or even to interlock to some shutdown system so that ventilation failure will shut off electrical apparatus in the surrounding area. In other words, the failure of the ventilation system itself needs to be considered, and the extent of the hazardous area without the ventilation present must be defined.

Since these circumstances will only be experienced under fault conditions, the resulting hazardous area will be zone 2.

Confidence of Calculated Results

Whilst calculations can produce some interesting results and may, at first sight appear to lead to more closely confined extents of zone, some careful evaluation should be undertaken before relying too closely on the result.

a) The equations require specific information on pressure, temperature, leak sizes and so on. Usually, there is significant variability of these properties and so it is always worthwhile varying one or two parameters, such as temperature and leak size, or plotting a graph of the results with varied properties to understand more closely what is happening.

b) Many of the equations make assumptions on the properties of gases and vapours. For example, where the ratio of specific heats is required in the original equations, the value has been taken as 1.3 and included as part of the numerical constant. (Ratio of specific heats can vary between 1 and 1.6)

c) Even if the complete rather than the simplified equations are used, there are still some assumptions. For example the assumption is made that the wind speed is 2m.s^{-1} (approx 4 mph or Beaufort Force 2). This is the lowest speed at which the equations hold true. For greater wind speeds the hazard will disperse in a shorter distance.

d) In all cases, the assumption is made that the hazardous area is 'well ventilated'.

e) In many cases the equations which are used are not applicable to vertical extents of zones.

Mixtures of Hazardous Gases

Sometimes a process includes mixtures of flammable gases. The extent of any zone clearly needs to be evaluated for the overall mixture.

This evaluation can be difficult since some properties, such as the vapour pressure of the combination, are not readily ascertainable. In all cases expert chemical advice should be obtained.

It should also be remembered that some mixtures will naturally separate out into their constituent parts. If this process will happen quickly, then individual gases may have to be considered separately. However, if the separation will take several days, as would be the case with say a small amount of hydrogen sulphide mixed with methane in a sewage treatment plant, then any release will disperse long before separation takes place, and the mixture can be considered as a whole. Clearly these aspects can make a significant difference since, for example, Methane has a relative vapour density of .55 (significantly lighter than air) whereas hydrogen sulphide has a relative vapour density of 1.19 (slightly heavier than air). In its mixture with methane, however, the hydrogen sulphide will tend to flow upwards.

Some reasonable guide to the LEL of gas mixtures may be obtained from the equation:

$$\text{LEL }_{\text{RESULTANT GAS MIX}} = $$

$$(100)/[(P_1/N_1)+(P_2/N_2)+ \ldots (P_X/N_X)]$$

Where:
P_1 etc. = proportion of the flammable content of the gas. (That is excluding any air or inert gas)
N_1 etc. = Vol % LEL of individual gases.

For example, a mixture containing the following gases may be dealt with thus:

Methane 80%   LEL 5.32%
Ethane   15%   LEL 3.22%
Propane 4%   LEL 2.37%
Butane   1%   LEL 1.85%

$$LEL_{RESULT} = (100)/[(80/5.32)+(15/3.22)+(4/2.37)+(1/1.85)]$$
$$= 4.55\%$$

The procedure is fully explained in BS 5345: Part 1: 1989.

Gas Detectors

The use of gas detectors is often considered in conjunction with area classification.

It must be understood that the intention of area classification is to provide an acceptable level of safety from ignition **if the hazard exists.** Gas detectors may be used to detect levels of concentration at which an alarm should be sounded, or a level at which the electrical apparatus should be switched off: a shutdown level. This activity does not, however, change the position that the electrical apparatus should be suitable for the hazard in which it is to be used.

Gas detectors are often not especially reliable or accurate (although in recent years there have undoubtedly been some major improvements) and it is not considered acceptable to rely on the gas detector to inform of an impending hazardous situation and then shut off the electrical apparatus. In any case, the release may occur so suddenly, especially under fault conditions that the response time of the detector and shutdown system would be too slow to ensure a shut down before ignition had taken place.

So gas detectors should be used to supplement the other measures, not replace them. Gas detectors also have a role in indicating levels of safety for operating personnel, and for local checks in a specific area before issuing a 'gas free certificate' or 'hot work permit'.

Organisation for the Classification Study

It will by now be clear that it is not practical for any one person to carry out an area classification study. Because the results of the study are primarily of interest to the Electrical and Instrument Engineers, and since the subject is placed within the Electrical Codes of practice, it often falls to these disciplines to carry out area classification work.

To attempt such work without the detailed process and chemical knowledge required is likely to be extremely unwise, and may lead to significant errors in the result.

The approach which seems best is to organise an area classification team or working group which will carry out the work and the necessary periodic reviews. Although this group may be headed by the Site Electrical Engineer it should comprise senior personnel from the Chemical Engineering and Process Engineering disciplines. Mechanical Engineers and Plant Engineers may also be called upon to give guidance on certain matters. In general, a small group (say three people) seems to work best, with others co-opted as necessary.

Area Classification Report and Results

Once the classification study has been completed, the results should be formally written up and agreed by all those involved with the classification work. The report should include an outline of the methods used and the reasoned argument for any decision which might prove controversial.

In addition to the written report, there should be a plan diagram of the site with the zones marked. In addition to the zone information, the diagram should include details of the Gas Group and Temperature Category applicable to the area.

The diagram should be readily available to all personnel who might reasonably require the zoning information. This will usually include electrical personnel, instrument personnel and anyone expected to specify, install or maintain any electrical apparatus within the area.

Reviews

Very few industrial plants remain unaltered for very long and even on new sites modifications to the processes or changes to the procedures and products produced will mean that changes are made which can affect the area classification.

There should be some system introduced which ensures that changes cannot be implemented without the knowledge of those responsible for the area classification of the site.

Regardless of any changes notified, the area classification study should be formally reviewed on a regular basis (say every one or two years) to ensure that no alteration has taken place which has escaped the attention of those responsible. Such reviews should also ensure that the classification still accords with current thinking on hazardous areas, and is consistent with any changes to the Code of practice or other guidance which have occurred in the intervening period.

Personnel and Education

New personnel on site who are likely to need the area classification information should be advised of the hazardous area and zone information at an early stage.

Clearly, if there are changes to personnel who comprise the area classification study group, then the new person taking up any of those key positions should be fully advised of the procedures which apply and the responsibilities involved.

40

Summary of Terminology Used for Area Classification

Absolute Temperature Equations nearly always require temperature in degrees Kelvin (K). Add 273 to °C

Auto Ignition Temperature See Ignition Temperature

Flash Point The lowest temperature at which a substance gives off sufficient vapour to produce a momentary flash when a flame is applied. (Not to be confused with Ignition Temperature)

Ignition Energy See Minimum Ignition Energy

Ignition Temperature Also known as Auto Ignition Temperature or Spontaneous Ignition Temperature. This is the temperature at which a substance will ignite due to the heating effect, even if there is no external source of ignition.

Minimum Ignition Energy The lowest energy which will ignite a mixture of the hazard and air. This value will exist somewhere between the UEL and LEL for a substance. For Hydrogen, the worst case, the MIE is 20 micro Joules.

Molecular Weight Also known as formula weight. Calculate by adding up the weights of each constituent.
The result is normally rounded to the nearest whole number. Hydrogen, although on its own must be treated as H_2 and cannot exist as H_1, takes its atomic weight (approx 1) when calculating molecular weights for other compounds.

Ratio of Specific Heats

For gases, the two most important specific heat capacities are that measured at constant pressure (C_P), and that measured at constant volume (C_V). C_P is greater than C_V, because when a gas is heated at constant pressure it has to do work against the surroundings in expanding. The ratio C_P/C_V is usually denoted by γ and varies from 1.66 for monatomic gases to just over 1 for more complex molecules. Where the ratio of specific heats is needed for area classification equations, a value of 1.3 can be taken as representative, and this value is normally included as a constant in the equation.

Vapour Density

Also known as Relative Vapour Density. This defines if a vapour is heavier or lighter than air, and thus is a number quoted with respect to Air = 1. Take care that the value obtained from tables is the comparison with air, and not a comparison with Hydrogen which is sometimes quoted in Chemical Handbooks. If no data is available, use molecular weight divided by 29.

Vapour Pressure

Also known as Saturated Vapour Pressure. This is needed for many equations used to determine the LEL. Since vapour pressure is dependent on temperature, the value usually has to be calculated from a partially complete equation which is available in tabular form in chemical handbooks. Values are often in units of mm Hg (Torr). To convert to Pa (= Nm^{-2}), multiply by 133.33.

Notes and References

1 The notable exception to this is intrinsic safety, where, although safe enough even for zone 0 use, ('ia') there are often significant cost savings in its selection, use and maintenance. These aspects are discussed in Section 3 of this book.

2 References and notes for table:

    i Masterton, W.L.; Slowinski, E.J.; Stanitski, C.L. Chemical Principles, Sixth Edition, Saunders College Publishing, 1986

    ii ACGIH Value. American Conference of Governmental Industrial Hygienists. Level at which workers may be exposed for 8 hour day, 40 hour week.

    iii OSHA ceiling value which must not be exceeded for more than 10 mins.

    iv LEL and UEL are more normally expressed as % concentration: in this case 4% and 46% respectively.

3 Justification and evolution of the equations is beyond the scope of this book. For those interested in the subject, the following sources may be found helpful:

    Perry, R.H. and Green, D. Perry's Chemical Engineering Handbook, 6th Edition, 1973, McGraw Hill.

    Dean, J.A. Lange's Handbook of Chemistry, 13th Edition, 1985, McGraw Hill.

    Tortoiseshell, G. Haztech Symposium 1987. The Scientific Approach to Area Classification.

4 At the time of writing there are a number of groups including a CENELEC working party, considering area classification with a view to a more comprehensive document (Standard). It has been suggested that, under certain circumstances, the presence of ventilation which is significant in quantity and integrity may be regarded as changing the zone number. It will be interesting to see if this proposal is actually adopted, and if so, how it will affect the practical results of area classification work.

5 The following documents and publications may serve as useful further reading for those requiring more information on area classification.

BS 5345

Part 1: 1989 General Requirements

Code of practice for the Selection, installation and maintenance of electrical apparatus for use in potentially explosive atmospheres.

This part of the Standard gives data on the more common chemicals and compounds, together with a wealth of other useful information on mixtures of compounds.

Part 2: 1983 Area Classification

Gives good general information on area classification. It falls short, however, of giving the detailed guidance needed for most situations.

ICI/RoSPA: 1972
Code on Electrical Installations in Flammable Atmospheres

An excellent document, but sadly no longer in print. It includes a comprehensive list of chemicals and compounds, and gives numerous examples of classification which can often be adapted to other situations.

Institute of Petroleum Code

The latest edition, published in 1990, includes more data than the earlier edition, and is well worthwhile. The examples given for classification of many common situations experienced in the petrochemical industry have been well thought out and save a lot of time and trouble.

IEC 79-10

Electrical apparatus for explosive gas atmospheres, Part 10: classification of hazardous areas.

BS 5345: Part 2 closely reflects this document, but there are some differences.

Lange's Handbook of Chemistry, 13th Edition. DEAN, J.A. Published by McGraw Hill

Although this book is expensive, it contains a wealth of information on properties of chemicals and is probably worth a place on the company bookshelf.

CHAPTER 3

Summary of Different Methods of Protection

Summary of Different Methods of Protection

Ex n, Ex N: N-Type Protection

Suitable for use in zone 2 only.

USA near equivalent: Non Incendive Protection.

Code of practice in UK: BS 5345: Part 7

IEC Standard: IEC 79-15

CENELEC Standard: none

British Standards: BS 6941   (Apparatus code Ex N)
                   BS 4683: Part 3 (Apparatus code Ex N or n)

Other Standards: BASEEFA Specification Memorandum Ex n

Live maintenance not normally permitted.

Main Requirements, Principles and Uses

Relies on good construction: environmental protection to at least IP54. (See Appendix 7). Mechanical strength tested by 3.5 Newton metre impact test. (7 Nm in some instances)

Requires non-sparking in normal use, proper cable entries and glands, terminals suitable for the current and environmental conditions.

Often used for lighting which is away from the worst of the hazard even if its job is to illuminate the hazard, junction boxes, rotating machines.

Cannot be 'EEx' certified, cannot bear the Distinctive Community Mark.

Requirements for installation quite onerous: sealing of cable glands, type of cable, etc.

No live maintenance without gas free certificate.

Is not the subject of a CENELEC Standard. The protection method is covered in BS 4683: Part 3 and the BASEEFA specification memorandum for N-type construction. This Standard has now been withdrawn, and is superseded by British Standard BS 6941: 1988, entitled 'Electrical apparatus for explosive atmospheres with type of protection N'. BS 6941 is not, however, a CENELEC Standard.

The foreword to BS 6941 states that it takes account of IEC 79-15.

For more details refer to Chapter 6

Ex d: Flameproof Protection

Suitable for use in zones 1 and 2.

USA near equivalent: Explosion Proof Protection.

Code of practice in UK: BS 5345: Part 3

IEC Standard: IEC 79-1

CENELEC Standard: EN 50 018   (Apparatus code EEx d)

British Standards: BS 5501: Part 5   (Apparatus code EEx d)
                        BS 4683: Part 2   (Apparatus code Ex d)
                        BS 229   (Apparatus code FLP)
                        BS 5000: Part 17   (Rotating machines)

If certified to CENELEC Standards by EEC approved body, may bear Distinctive Community Mark.

Live maintenance not permitted.

Main Requirements, Principles and Uses

Relies on construction to ensure that any ignition of the hazard inside the enclosure will not transmit to the atmosphere outside the enclosure.

Generally requires certification for each specific application, although there are a number of general approvals which allow the user to put a variety of apparatus inside a certified enclosure without invalidating the certificate, e.g. combinations of terminals, zener barriers, relays.

May be suitable for all Gas Groups, although for Group IIC special requirements are needed usually resulting in boxes with screw tops, etc.

Onerous requirements for installation and maintenance as it is this aspect which makes or breaks the safety of the apparatus.

Maximum gap between lid and body of box specified. Note: flameproof boxes are not necessarily waterproof. The flange (gap) may be sealed with non-setting grease to increase weather protection. (Beware the use of silicone grease which may affect gas detectors in the vicinity!).

The technique has been heavily criticised offshore because experience has shown that flameproof boxes often fill up with sea water and corrode badly. If greater sealing than greasing of the flanged joint is required, then the enclosure must be designed with a sealing gasket. The presence of such a gasket on a correctly designed enclosure is purely for environmental protection and plays no part in the security of the method of protection. Gaskets, etc., must not be added to enclosures which have not been designed and certified to include them. In North America, the environmental protection of an 'Explosionproof' enclosure may be included as part of the protection code, using the NEMA Code System (Appendix 7).

Requires correct cable glands to be inserted into the box (usually pre-threaded by the manufacturer), cable to comply with the Code of practice. Correct sealing of unused cable entries required.

There are two conventional ways of making cable entries to a flameproof enclosure:

Direct Cable Entry: via a correctly approved and fitted flameproof cable gland

Indirect Cable Entry :   via a separate compartment on the side of the enclosure.

Indirect cable entries frequently employ increased safety ('e') techniques. So there may be an Ex e terminal compartment attached to the Ex d enclosure. In this case the coding for the apparatus would read EEx de .... indicating a primary method of protection 'd' with some increased safety aspects as well.

For more detail refer to Chapter 9.

Ex p: Pressurization
Suitable for use in zone 1 and 2.

Code of practice in UK: BS 5345: Part 5

IEC Standard: IEC 79-2

CENELEC Standard: EN 50 016   (Apparatus code EEx p)

British Standard: BS 5501: Part 3   (Apparatus code EEx p)

If certified to CENELEC Standards by EEC approved body, may bear Distinctive Community Mark.

Live maintenance not permitted.

Main Requirements, Principles and Uses

Relies on creating a non-hazardous (safe) area inside the enclosure so that ordinary (non-certified) apparatus may be used inside the enclosure with few, if any, restrictions.

In order to attain the required degree of safety it is necessary to first purge the enclosure by opening a vent and passing clean air or inert gas through the enclosure, then shutting off the vent and over pressurizing with clean air or inert gas, such that the pressure inside the enclosure is above the ambient pressure outside. The enclosure will have some monitoring electronics attached to ensure that these operations have been carried out before the apparatus inside the enclosure is powered up, and to shut off the power if the pressure fails. This monitoring apparatus usually encompasses other methods of protection: for example, intrinsically safe pressure sensors operating via zener barriers located in a flameproof box bolted onto the side of the purged container.

The technique is often used for control panels which must be located in a hazardous area, or which, although normally in a safe area, must be capable of safe operation in the event that the control room is invaded with the hazard.

Also used for locating apparatus which is not, or cannot be otherwise certified for use in the hazardous area (e.g. VDU's).

Tends to be very expensive. Probably should only be used if there is no other practical solution.

In the USA there are three grades of purging:

Type Z: reduces the classification within an enclosure from Division 2 to Non-hazardous

Type Y: reduces the classification within an enclosure from Division 1 to Division 2

Type X: reduces the classification within an enclosure from Division 1 to Non-hazardous.

For more detail refer to Chapter 10.

Ex e: Increased Safety

Suitable for use in zones 1 and 2.

Code of practice in UK: BS 5345: Part 6

IEC Standard: IEC 79-7

CENELEC Standard: EN 50 019   (Apparatus code EEx e)

British Standard: BS 5501: Part 6   (Apparatus code EEx e)
                        BS 4683: Part 4   (Apparatus code Ex e)

If certified to CENELEC Standards by EEC approved body, may bear Distinctive Community Mark.

Live maintenance not permitted.

Main Requirements, Principles and Uses

Possibly increasing in popularity, the technique relies on constructional safeguards to ensure that the apparatus does not contain normally arcing or sparking devices, or hot surfaces that might cause ignition. Measures are applied to reduce the possibility of failure, and hence arcing or sparking of the normally non-sparking parts. This is achieved by:

- the use of high integrity insulation

- the temperature de-rating of insulation materials

- enhanced creepage and clearance distances

- particular attention to terminal design

- protection against the ingress of solids and liquids

- impact test requirements for the enclosure

- control of maximum temperatures (e.g. especially for motors, where a safety cut out may be required)

- requirements for air gaps and running temperatures (for motors).

The technique is not applicable where rated voltages exceed 11 kV.

Increased Safety is often used for rotating machines and, in, recent years has become popular for terminal boxes and terminal assemblies because, like flameproof protection, increased safety is acceptable for zone 1 use but the costs of an increased safety terminal box are much lower than for a corresponding flameproof box.

Cable glands to increased safety apparatus should either be of component certified type or, at the user's discretion and responsibility, may be other industrial cable glands which afford IP54 protection, will clamp the cable sufficiently to withstand the required pull test, and are mechanically strong.

For more detail refer to Chapter 12.

Ex o: Oil Immersion

Suitable for use in zone 2 only (UK), but zones 1 and 2 in some countries.

Code of practice in UK: BS 5345: Part 9 (Not yet published)

IEC Standard: IEC 79-6

CENELEC Standard: EN 50 015   (Apparatus code EEx o)

British Standard: BS 5501: Part 2   (Apparatus code EEx o)

If certified to CENELEC Standards by EEC approved body, may bear Distinctive Community Mark.

Live maintenance not permitted.

Main Requirements, Principles and Uses

The technique is most commonly applied to heavy current switchgear and transformers, but recently has found application for the remote enclosures of distributed control systems and other low power and normally non-sparking apparatus.

The filling medium is mineral insulating oil. Proposed amendments to the CENELEC Standards will allow use of other fluids.

Requirements for the enclosure are onerous, but the end result is a sound and robust piece of apparatus. Both 'sealed enclosure' and open to top up reservoir enclosures are permitted. There must always be some means of checking the oil level.

Sealed enclosures require a pressure relief valve, but the design will be such that this will only operate under major failure conditions.

For more detail refer to Chapter 8.

Ex q: Sand or Powder Filling

Suitable for use in zone 2 only (UK), but zones 1 and 2 in some countries.

Code of practice in UK: BS 5345: Part 9 (Not yet published)

IEC Standard: IEC 79-5

CENELEC Standard: EN 50 017   (Apparatus code EEx q)

British Standard: BS 5501: Part 4   (Apparatus code EEx q)

If certified to CENELEC Standards by EEC approved body, may bear Distinctive Community Mark.

Live maintenance not permitted.

Main Requirements, Principles and Uses

The technique is most commonly found in West German manufactured and Certified apparatus. It is not commonly used in the UK. Quite often, the filling medium is very small glass balls.

The enclosure is normally factory sealed, although provision is made in the Standard for top up opening designs.

Most common applications are for electronic assemblies in telephone units, and power supplies for use in the hazardous area where the supply has an intrinsically safe output.

For more detail refer to Chapter 7.

Ex m: Encapsulation

Acceptable for use in zones 1 and 2.

Code of practice in UK: Not yet published

IEC Standard: None

CENELEC Standard: EN 50 028   (Apparatus code EEx m)

British Standard: BS 5501: Part 8   (Apparatus code EEx m)

If certified to CENELEC Standards by EEC approved body, may bear Distinctive Community Mark.

Live maintenance not permitted.

Main Requirements, Principles and Uses

A relatively new method of protection (the CENELEC Standard was published in 1988) it is not yet clear what the normal uses will be. However, since the encapsulation means that apparatus is not readily repairable, it will probably be mostly used for low cost 'throw away' items. Much apparatus which would otherwise have used special protection ('s') will now be able to gain full EEx certification using this technique.

For more detail refer to Chapter 11.

Ex s: Special Protection

Normally suitable for use in zones 1 and 2. Possibility of use in zone 0 if the certification documents specifically allow it.

Code of practice in UK: BS 5345: Part 8

IEC Standard: None

CENELEC Standard: None

British Standard: None

Other Standards: BASEEFA SFA 3009 (Apparatus code Ex s)

Cannot bear the Distinctive Community Mark.

Live maintenance not permitted. (Unless specifically mentioned in certification documents)

Main Requirements, Principles and Uses

The idea of Special Protection is to encompass designs which are considered by the Testing and Certification Authority to be safe for use in a hazardous area, but which do not comply with any of the recognised methods of protection. It may also apply to gas detectors and flameproof enclosures with breathing devices which, at the time of writing are not covered within the flameproof Standard.

The apparatus will normally be constructed so as to meet the requirements of BS 5501: Part 1 (EN 50 014) and, depending upon the intended use of the apparatus, additional requirements may apply for Drop and Impact Tests, Sealing and Encapsulation.

For more detail refer to Chapter 11.

Ex ia and Ex ib: Intrinsic Safety

Suitable for use in zones 1 and 2 (Ex ib)
Suitable for use in zones 0 and 1 and 2 (Ex ia)

In the USA there is only one grade of Intrinsic Safety. suitable for both Division 1 and 2. Canada, however, recognises the 'ia' and 'ib' distinction.

Code of practice in UK: BS 5345: Part 4

IEC Standard: IEC 79-11

CENELEC Standard: EN 50 020   (Apparatus Standard)
                                              (Apparatus code EEx ia or ib)
                             EN 50 039   (System Standard)

British Standard: BS 5501: Part 7 (Apparatus Standard)
                                              (Apparatus code EEx ia or ib)
                             BS 5501: Part 9 (Systems Standard)
                             BS 1259   (Apparatus code Ex ia or ib)

Other Standards: BASEEFA SFA 3012 (Apparatus code Ex ia or ib)
                           BASEEFA SFA 3004   (Apparatus code [Ex ia or ib])

If certified to CENELEC Standards by EEC approved body, may bear Distinctive Community Mark.

Live maintenance permitted.

Main Requirements, Principles and Uses

Intrinsic safety is the only method of protection which does not rely on mechanical integrity to ensure safety from causing ignition. The technique of intrinsic safety is the highest integrity method of protection available. It is, in practice, the only method suitable for use in zone 0 (Ex ia only).

The concept is to design and construct the apparatus in such a manner that the electrical energy available from the circuit will be less than the minimum ignition energy of the hazardous gas into which it is to be placed.

Minimum ignition levels are established by the use of the Break Flash Apparatus, or by reference to the minimum ignition curves which are published in the Standards.

Requires some defined safety interface between the safe and hazardous area so that the maximum power into the intrinsically safe circuit in the hazardous area can be established.

The two grades of intrinsic safety 'ia' and 'ib' refer to the safety of the apparatus or system with various faults applied.

It should be remembered that Intrinsic Safety is a **system concept.** Safety cannot necessarily be assumed just by reference to one item of apparatus.

For more detail refer to Chapter 13.

METHOD OF PROTECTION	Ex CODE LETTER	CENELEC CODE	ZONE OF USE—UK			CODE OF PRACTICE	BRITISH STANDARD	CENELEC EN 50	NORMAL USE
			0	1	2				
N-TYPE (NON-INCENDIVE)	N	N/A			√	BS-5345 Part 7	BS-6941	None	Lighting Motors Junction Boxes
INCREASED SAFETY	e	EEx e		√	√	BS-5345 Part 6	BS-5501 Part 6	019	Rotating machines Junction Boxes
FLAMEPROOF	d	EEx d		√	√	BS-5345 Part 3	BS-5501 Part 5	018	Motors, Junction boxes. Lighting, switches instrumentation
PURGE/ PRESSURISED	P	EEx p		√	√	BS-5345 Part 5	BS-5501 Part 3	016	Control panels. Usually apparatus which cannot be protected by other techniques
SPECIAL PROTECTION	s	N/A	(√)	√	√	BS-5345 Part 8	None	None	If tested and shown safe but not complying with a recognised protection concept
ENCAPSULATION	m	EEx m		√	√	Not yet published	BS-5501 Part 8	028	Not yet established
OIL FILLING	o	EEx o			√	BS-5345 Part 9	BS-5501 Part 2	015	Switchgear and heavy current apparatus. Some recent use for distributed control systems
SAND/POWDER FILLING	q	EEx q			√	BS-5345 Part 9	BS-5501 Part 4	017	Power supplies—especially for weighing systems with I.S. outputs. Telephone electronics
INTRINSIC SAFETY	ib ia	EEx ib EEx ia	√	√ √	√ √	BS-5345 Part 4	BS-5501 Part 7	020	Instrumentation—measurement & control circuits. Lower power apparatus

NOTES

1 The CENELEC Standards are published as British Standards in the series BS 5501.

2 Previous Standards mentioned in the comments column are no longer used for Certification apart from some variations to old Certificates. New applications will be certified to the Standard shown. The continued use of old Certificates is not precluded and much apparatus is still marketed to the old Standards.

3 Unless stated, Certificates issued to the CENELEC Standards by the appropriate Certification Authority will meet the EEC Directives and thus such apparatus may use the Distinctive Community Mark.

SAFETY ACHIEVED BY	NOTES/COMMENTS
Non-sparking in normal operation	Not CENELEC certifiable and thus not included in EEC Directives. Original intention was to allow 'good quality industrial grade apparatus' to be used in Zone 2. Although usually certified, the Code of practice still allows apparatus for Zone 2 to be used without certification. Previous Standard BS-4683 Pt3.
Non-sparking in normal operation and under certain fault conditions	Because of restrictions on conductor size, etc., not really suitable for instrumentation. Previous Standard BS-4683 Part 4
Suppression of frame propagation. Any explosion within the enclosure does not ignite surrounding atmosphere	Correct installation and subsequent good maintenance essential. Very widely used. Most modern installations will use intrinsic safety for instrumentation, but flameproof instrumentation is still popular in some industries. Previous Standard BS-229, BS-4683 Part 2
Removing (purging) any hazardous atmosphere inside the enclosure and then maintaining the interior as a non-hazardous area	Useful for custom built apparatus. Often a component certified control system can be 'bolted-on' to a cabinet to achieve a pressurized enclosure with the minimum of Certification effort. Tends to be the two or so manufacturers of certified Ex p control systems who arrange the overall Certification.
Often involves the use of some encapsulation—thus preventing atmosphere reaching arcing/sparking parts	Not CENELEC certifiable and thus not included in the EEC Directives. Although technically may be suitable for Zone 0 use, it is almost always restricted to Zones 1 & 2 use, Read the certificate carefully to establish conditions of use and special requirements for installation etc. Standard for certification is SFA-3009 (BASEEFA).
Potting or encapsulating apparatus in resin to stop hazard reaching electrical parts	CENELEC Standard published in 1988. Not yet included in the EEC Directives and at present any Certificates issued will be Certificates of Assurance and not Certificates of Conformity. Will to some extent supersede the technique of special protection.
Prevention of hazard or air at source of ignition	Not widely used. Advantages are environmental protection and improved electrical reliability due to oil dissipating any hot spots. Most other EEC countries allow EEX o for use in Zones 1 & 2. There may be a change in the UK Code to allow low power/non-sparking apparatus using this technique in Zone 1 if there is sufficient interest.
Electronics submerged in sand, and thus suppression of flame propagation	Not that widely used, but perhaps more popular in Germany. Most other EEC countries allow its use in both Zones 1 & 2. There may be a change in the UK to this effect in the future. Usually uses small glass or silica beads rather than sand or powder. The 'q' stands for quartz. Usually factory sealed units.
Maintaining electrical energy at levels below the minimum ignition energy for the hazard	The most popular technique for instrumentation circuits The only method of protection (ia) acceptable in Zone 0. In spite of its high level of protection there is considerable freedom for the use of non-certified 'simple apparatus' (non energy storing). Greater freedom on use of cables. Live maintenance permitted. (Note 4) Needs defined safety interface. (Note 5) Only technique allowing overall system certification. (Note 6). Previous Standards BS-1259, SFA-3012

4 Apart from the possibility of N-type protection in zone 2, intrinsic safety is the only method of protection where live maintenance is permitted.

5 Interface is usually a zener safety barrier or galvanic isolator. Galvanic isolation is a requirement for zone 0 applications in Germany.

6 Systems comprising several items of intrinsically safe apparatus may be certified to EN 50 039 (BS 5501: Part 9).

CHAPTER 4

Standards & Certification: Apparatus Design Assurance

Standards & Certification: Apparatus Design Assurance

The International Electrotechnical Commission (IEC) has produced a series of Standards dealing with electrical apparatus for hazardous areas. These documents are all numbered in the series 79-xx. (eg. 79-11 Electrical apparatus for explosive gas atmospheres, Part 11: Construction and test of intrinsically safe and associated apparatus.)

The Standards describe the various methods of protection, and make recommendations on the type (severity) of hazardous area in which they may be used.

This chapter is concerned with the Standards which apply to the design and certification (sometimes called approval) of apparatus to the different established methods of protection which exist.

Although the precise details of the different methods of protection may vary from country to country, the main recognised methods have basic rules which must be followed to construct apparatus which will conform to the protection method and will successfully pass the tests to which it may be subjected in the course of any certification process.

Certification is a process of assessing a design to the specification set down in a Standard. The Certification Authorities are not, in the main, the Standards writing bodies (although they often have a strong presence on those bodies) but are rather the independent service which gives a third party attestation that a design complies with the requirements of the Standard.

The main methods of protection for which Standards exist are:

Flameproof or explosionproof apparatus
Intrinsically safe apparatus
Increased safety apparatus

Purged/pressurized apparatus
N-Type and non-incendive apparatus
Oil filled apparatus
Sand filled apparatus
Encapsulated apparatus

Intrinsically safe systems

It should be noted that, with the exception of intrinsic safety, the Standards for certification only concern themselves with the apparatus design, not the way apparatus is connected or used.

The IEC Standards cover the prime requirements for each method of protection, but they do not give sufficient detail to ensure a uniform assessment, and so they are not normally used for certification purposes. Instead, different countries and organisations produce detailed Standards, based on the requirements and recommendations of the IEC Standards, to which apparatus may be fully assessed and certified.

With one or two exceptions, the process of third party certification is the only way that a manufacturer can be sure that his product will be safe for sale and use in hazardous areas, and the end user normally requires this assurance before he is prepared to make a purchase.

Although there are differences from country to country in the way in which the Standards and certification procedures work, the subject can be readily understood by an appreciation of a few main organisations.

Europe (CENELEC)

Within Europe, the Committee for Electrotechnical Standardisation (CENELEC) [1] has produced a series of Standards, which reflect the IEC recommendations. These Standards are listed in Appendix 5.

Because it is normal for members of CENELEC to publish the identical text of CENELEC Standards in their own language, there are corresponding National Standards in many of the CENELEC countries. For example in the UK, the British Standards Institution publishes the CENELEC Standards as British Standards in the English language.

Apparatus which is certified to the CENELEC Standards will bear the code EEx, followed by the code for the method(s) of protection, the gas group and temperature class.

> For example: EEx d IIB T6
>
> For an item of flameproof apparatus certified to the CENELEC Standards for use in gas group IIB and with a temperature class of T6.

Since the certification exercise is essentially a design approval, the onus is on the manufacturer to produce his product in accordance with the certification documents, and certification recorded drawings. Different Certification Authorities within Europe take a different view on whether or not they are required to be involved in the QA arrangements of the manufacturer, and hence have some knowledge of whether or not the company which holds a Certificate is likely to produce products which conform to the certification conditions.

At the time of writing, the whole concept of Quality Assurance Assessments and Audits for Certificate holders is under review within Europe, and there may well be a more unified approach in the next few years. At present, BASEEFA and Sira Certification Services (both UK Certification Authorities) both carry out some form of audit on the manufacturer of Certified apparatus, but this approach is not yet common to other countries' Certification Authorities.

65

BASEEFA Licensing Arrangements

Part of the CENELEC Standards requires that apparatus which is certified to the Standards is marked in the prescribed manner. This marking includes the mark of the Certification Authority. BASEEFA license the Certificate holder to reproduce their logo

Figure 4.1 *BASEEFA Logo*

on the apparatus label. The Licence, which looks very similar to the Certificate, expires every three years, and has to be renewed. BASEEFA often take this opportunity to carry out a Quality Audit on the Certificate holder.

In practice, this regular audit may not take place. There are too many Certificate holders, and not enough personnel at BASEEFA to carry out the audits, without seriously affecting their actual certification work. The position has improved recently, by the appointment of a Quality Assurance Officer within BASEEFA, but it is still not clear just what approach will be taken in the future. The audit is, in any case, only really concerned with ensuring good control over the revision and issuing of certified drawings, with the calibration procedures of any crucial measurement or test apparatus required for the manufacture of certified products, and with purchasing control and stores segregation and route identification in the manufacturing process. Thus, the BASEEFA audit does not attempt to be a full QA Audit or Assessment.

Other countries do not carry out this licensing procedure at all. The Certification Authority freely allows the use of their mark on any Certified product. However, in view of some recent situations where certified products have been found after installation at the end user, not to comply with the certification requirements, it has been suggested that a closer control is required over manufacturers. Various organisations, for example representing the offshore industry, have undertaken to carry out independent assessments on their key suppliers, and this type of activity may eventually be linked with the certification procedure.

There is, in any case, a requirement on the BASEEFA application form to comply with a Quality system which is equal to BS 5750: Part 2. (ISO 9002) [2]

EEC

Within the European Common Market, EEC, the CENELEC Standards have a special significance, since they have been formally adopted by the EEC. A number of European Directives call up the CENELEC Standards and state that apparatus which conforms to those Standards and has been assessed as such by a listed Test House of a member country may bear the Distinctive Community Mark.

Figure 4.2   *EEC Distinctive Community Mark*

This mark implies that apparatus meets the minimum EEC requirements, and thus does not require re-testing when transferred from one member country to another. Since it only specifies a minimum level, some countries specify additional requirements, and this can appear as a trade barrier between member countries.

It should be understood that the Distinctive Community Mark is not a Certification Mark, but rather a trading mark between member countries of the EEC. It is possible that the current mark will be replaced in the near future with a new EEC mark, but the details are not yet clear, and in any case there has been considerable opposition to the use of a new mark since it will add confusion and will not be peculiar to electrical apparatus in hazardous areas.

Unfortunately, because the EEC official bodies tend to move rather slowly, there is often a significant time lapse between the publication or amendment to a CENELEC Standard, and its incorporation to the EEC Directive. Thus it has become normal practice for EEC member countries to issue Interim Guides to the CENELEC Standards to allow them to use the latest edition of the Standards before the EEC have formally updated the relevant Directives.

Individual Countries Within Europe

Although most applicants within Europe will automatically now request Certification to the CENELEC Standards, most countries still have some National Standards in force. This may occur where a method of protection exists which is not yet incorporated into the CENELEC Standards, as for example, with N-Type protection. In such cases certification can be obtained to the appropriate National Standard, and will bear the code 'Ex' again followed by the rest of the certification information.

In some cases, an applicant for certification might find that he cannot meet the stringent requirements of the CENELEC Standards, but can meet the slightly more flexible requirements of an earlier National Standard. He is free to apply for such certification, which again will be certified with an Ex code.

Since, with the exception of Component Certificates (which are normally valid for three years unless renewed), certificates are valid for all time, many manufacturers hold quite old certificates which are still being used for current manufacture. This means that it will be many years before all Standards prior to the CENELEC Standards become truly obsolete. For example, many UK manufacturers still sell flameproof apparatus to the old British Standard BS 229 which will bear the mark shown in Figure 4.3. This Standard is, however, no longer used for certification, and indeed is no longer available from BSI.

Figure 4.3 *Mark Indicating Apparatus Certified to BS 229*

Further examples of marking codes and details of the Certification Standards for the different methods of protection are given in the Appendices 1, 5 and 6.

Types of Certification Document

There are various types of certificate, depending on the Standard to which the apparatus has been certified, and the status of the certification. Within Europe, the main types of certificates which will be encountered are as follows.

COMPONENT CERTIFICATE

This is not full certification, and the apparatus to which the Component Certificate refers will need some additional protection before it can be used in the hazardous area. Examples of Component Certificates are a flameproof enclosure without any contents, an intrinsic safety opto-isolator, terminals for use in an increased safety enclosure, etc.

The Component Certificate can be readily identified because the Certificate number will end with the letter 'U'.

There are reduced marking requirements applicable to component certified items, because they are very often too small to allow all the normal labelling. However, in all cases at least the Component Certificate number should be present.

CERTIFICATE OF ASSURANCE

Apparatus certified to a National (not CENELEC) Standard, or certified to a CENELEC Standard by a Certification Authority outside the EEC or if the Certification Authority is within the EEC, but the CENELEC Standard has not been included in the EEC Directives.

The certification code will be Ex if a National Standard applies, and EEx if a CENELEC Standard applies.

CERTIFICATE OF CONFORMITY

Applies to apparatus certified by an EEC Certification Authority to a CENELEC Standard recognised by the EEC Directives.

SYSTEM CERTIFICATE (Intrinsic Safety Only)

A certificate covering the interconnection of several items of apparatus. This type of certificate is only awarded where intrinsic safety is involved, since intrinsic safety is the only method of protection which has Standards covering both apparatus and systems of interconnected apparatus.

CERTIFICATES WHERE SPECIAL CONDITIONS OF USE APPLY

Where there are special conditions of use for certified apparatus, for example if it requires unusual earthing arrangements to maintain the required level of safety, then the certificate number will end with the letter 'X'. In the past, the letter 'B' has been used instead of 'X', and may still be encountered on some older certificates.

LICENCE

Only issued by BASEEFA, to authorise the Certificate holder to reproduce the BASEEFA logo on his label, and to indicate that Quality Assurance Audits have been acceptable. It is a good idea to avoid sending copies of the Licence to end users who request copies of Certificates, since unless the document is read carefully, it may lead to the end user believing that a Certificate is about to expire.

North America

In North America, the Certification scene is slightly less well defined than within Europe because the IEC recommendations for hazardous area classification and methods of protection are not followed. Some movement towards the IEC system has been made in recent years, but at present the old terminology of Divisions, and the use of the Class I gases and vapours, together with terms such as 'explosionproof' and 'non-incendive', mean that two complete sets of terminology need to be understood.

USA

In the USA, although there are a comprehensive series of National Standards, published either by the National Fire Protection Association, NFPA or the American National Standards Institute, ANSI, there is no National Certification Authority such as PTB or BASEEFA. Instead, there are two independent Testing Authorities; Factory Mutual Research Corporation (FM), and Underwriters Laboratories (UL).

Figure 4.4 *Marks of Factory Mutual and Underwriters Laboratories*

Each of these testing authorities will carry out testing and evaluations to either the National Standards or to their own published Standards. The main requirements for all methods of protection are based on the National Electrical Code, NEC 70, published by NFPA, and re-issued every three years.

In the USA, the term Approval is preferred to Certification. UL approvals are called listings, because the approved design appears in UL's published list of apparatus.

Both Factory Mutual and Underwriters Laboratories carry out some form of audit on manufacturers who hold approvals, and expect to see evidence of correct drawing control and manufacturing procedures of the approved products.

Details of applicable USA Standards and labelling will be found in Appendices 1 and 5.

Canada

In Canada, the Canadian Electrical Code Part 1 defines the requirements for electrical apparatus in hazardous areas. The Canadian Standards Association (CSA) publish various Standards for the different protection concepts, and also offer an approval service through their Test Laboratories in Toronto. Apparatus which has been approved by the CSA will bear their mark. CSA carry out routine audits of manufacturers who hold CSA approvals. Audits of manufacturers outside Canada are normally carried out by a CSA agency located in that country.

Figure 4.5   *CSA Mark*

Underwriters Laboratories of Canada, ULC are also recognised.

There is no national code of practice specifically dedicated to electrical apparatus for hazardous areas within Canada, and the requirements for installation of such apparatus is governed on a provincial rather than a national basis. The inspection branch of each province and territory have a major influence on the type and installation of apparatus on, for example, petrochemical plants. To ensure that there is a general similarity between each area, the inspection branches meet on a regular basis to formulate their approach. Thus, in reality there is similar guidance throughout the country.

Although Canada follows the classification and terminology of North America, it has close links with the IEC and CENELEC system. There is in fact a reciprocal arrangement between Canada and the UK, and holders of BASEEFA Certification can usually obtain a CSA approval by submitting their certificates and test report to the Technical Help to Exporters branch of British Standards, where there is a CSA agency which issues CSA approvals on the basis of BASEEFA's work. However, this arrangement does not always work, and there have been instances where complex applications have been refused by BSI and referred to Canada.

BASEEFA and CSA aim to accept each other's test results, so even if an application has to be restarted, there is usually some advantage in submitting the Test Report issued by BASEEFA or CSA as the case may be. Since there are minor differences in the Standards, sometimes the tests required are not identical, and the arrangement can appear to be unworkable. It is, however, a step in the right direction.

Australia

In Australia, the Standards Association of Australia SAA produces National Standards, and offers a certification service. The main requirements are very similar to the European Standards, and indeed there are some reciprocal arrangements for acceptance of test results between European bodies and the SAA. For some methods of protection (notably intrinsic safety) certification with a European Certification body such as BASEEFA should allow automatic SAA approval without further assessment.

South Africa

In South Africa, the South African Bureau of Standards SABS both publish Standards and offer an approvals service. The Standards are very closely linked with the pre CENELEC Standards used in the UK. For example SABS Standard 549-1977 closely reflects BASEEFA document SFA 3012.

Figure 4.6 *Mark of the South African Bureau of Standards*

Certification Procedure

Although the precise requirements for obtaining Certification vary from Authority to Authority, there are a few simple rules which should stand the applicant in good stead in all cases.

Preparing the Design

The first stage in obtaining Certification is to prepare a design which will, at least broadly, conform to the chosen Standard. If there is doubt over which Standard is the most suitable, the certification authorities will usually give some guidance before an application is made.

Before any detailed design work is carried out, the person responsible must make sure that they are in possession of the latest revision of the Standard. This is especially important where the European Standards are concerned, since there are frequent amendments and interpretation sheets which can have very significant bearing on detailed design aspects.

It is worth checking at the earliest stage if the apparatus needs to be certified at all. Especially where the design is to utilize intrinsic safety, it may be possible to make use of the 'simple apparatus' rule concerning non-energy storing apparatus. If these conditions [3] are met, and the apparatus can be connected via a conventional zener safety barrier, then certification may not be needed at all.

When preparing the design, keep in mind that the drawings and data will have to be assessed by someone who is not at all familiar with your product. Thus the drawings should be clear and concise, and only specify those aspects which are necessary to demonstrate compliance with the Standard. Do not be tempted to use manufacturing drawings; they are normally far too detailed, and will stipulate all kinds of aspects which are not necessary for Certification. This has the effect of firstly making the work of the Certification Officer much harder, because he has to sift through a lot of irrelevant data, and secondly because once a drawing is stamped as a Certified drawing, it cannot be changed without reference to the Certification Authority and an application for a Variation.

Once the design drawings are at a stage where they are ready to submit to the Certification Authority, the application can be made. [4] Some Authorities have a special application form (eg BASEEFA) whilst others purely need a letter making the formal application.

Where Certification to EEC/CENELEC Standards is sought, it should be remembered that an assurance from the manufacturer is required that the apparatus will not cause danger or be unsafe in situations other than the hazardous area aspect which is to be assessed. In addition, a statement should be made that the manufacturer operates a Quality Assurance system which is at least as good as ISO 9002 (BS 5750: Part 2)

Normally, an nominal application fee is required, and should accompany the application, which should include at least three copies of all relevant drawings, parts lists and other data. It is not normally advisable to send test samples at the application stage; these will be requested later on.

Once the application has been received by the Certification Authority, it will be given an initial assessment. This will lead to the applicant receiving a letter, normally within a few weeks of the submission, giving an estimate of the fees for the Certification work, requesting samples for test and any additional drawings, and enclosing an invoice for the estimated fee. This stage is crucial to the timing of the exercise, since the work does not normally enter the 'pickup queue' for testing and assessment, until everything requested by this letter - including payment of the invoice - has been received.

Once the Certification Authority has received all the information etc. requested, the work enters the queue for detailed testing and assessment. When it reaches the top of the queue, it will be assigned to the next available Certification Officer who is competent to handle the work. [5]

During the period of the detailed assessment, the Certification Officer will usually have a number of questions, and requests for additional information. Again, it is crucial to ensure that any such questions are promptly answered. If the applicant introduces long delays, the Officer will have picked up other work, and it may be some time before he is able to restart the application. [6]

Once the work has been successfully completed, and any additional drawings, data etc. provided, there will be a request for any final payment due (normally the amount of any overspend from the original estimate will have been advised earlier on) and once the payment has been received, the Certificates and Stamped Certified Drawings will be despatched.

Figure 4.7 shows the key stages in the certification process.

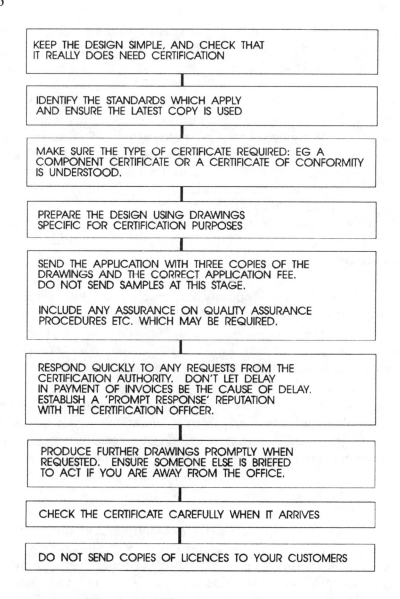

Figure 4.7   *Key Points in Certification Procedure*

Notes and References

1   The following countries are members of CENELEC:
    Austria
    Belgium
    Denmark
    Finland
    Germany
    Greece
    Ireland
    Italy
    Luxembourg
    Netherlands
    Norway
    Portugal
    Spain
    Sweden
    Switzerland
    United Kingdom

2   The ISO is the International Organisation for Standardization. Countries within ISO adopt the ISO Standards nationally.

3   The requirements for simple apparatus are explained in Chapter 13.

4   The procedure described in this section is based on the requirements of BASEEFA.

5   Most Certification Officers specialise in one particular technique, for example intrinsic safety or flameproof protection, and within that specialisation, some are more used to certain types of design. The Certification Authority usually tries to ensure that the work is directed to the most suitable person for evaluation.

6   Some Certification Authorities keep a log of all waiting time on an application, and close the application file if the cumulative waiting time exceeds six months.

CHAPTER 5

Dust Hazards

Dust Hazards

APPLICABLE STANDARDS	
BS 6467	Electrical Apparatus with protection by enclosure for use in the presence of combustible dusts
Part 1	Specification for apparatus (1985)
Part 2	Guide to selection, installation and maintenance
ISA-S12.10	Area classification in harzardous dust locations
BS ? [1]	Guide to the use of BS 5501 Electrical Apparatus in the presence of combustible dust

In the main, this book is concerned with the protection of electrical apparatus for use in flammable gases and vapours. However, dusts can also present an explosion hazard, and although the primary considerations of preventing any electrical apparatus located in the dusty environment from causing ignition are similar to those where the hazard is a gas or vapour, there are some additional aspects which need to be addressed.

Terminology for Dust Hazards

Dusts are defined as small solid particles that settle out under their own weight but that may remain suspended for some time. If such dust is combustible or ignitable in mixtures with air, it is a combustible dust.

The area classification procedures and terminology explained in Chapter 2 do not apply to dust hazards. Dusts are classified in terms of their ignition temperature, and the likelihood of the hazard existing is defined by the zone, but the zone definition is different from that previously introduced.

There are two zones for dust hazards: [2]

Zone Z

An area in which combustible dust is, or may be, present as a cloud during normal processing, handling or cleaning operations in sufficient quantity to be capable of producing an explosible concentration of combustible or ignitable dust in a mixture of air. [3]

Zone Y

Areas not classified as Zone Z in which accumulations or layers of combustible or ignitable dust may be present under abnormal conditions and give rise to ignitable mixtures of dust and air. [3]

The extent of the zone is the distance in any direction from a source of release to the point where the hazard associated with that zone is no longer considered to exist.

As for gas and vapour hazards, catastrophic failures are not to be considered in ascertaining possible sources of release, since such occurrences would give rise to zones of excessive extent, and the normal criteria for plant design, inspection and maintenance should have ensured that such failures (for example the fracture of a pipeline or process vessel) are extremely unlikely.

Areas which are neither Zone Z nor Zone Y are non-hazardous areas, and in these locations, no special precautions are required for prevention of electrical ignition of the dust hazard.

Main Safety Criteria

The main risk of ignition of dust hazards is from hot surfaces. Dusts may settle on surfaces and the build up can give rise to a concentration which could be ignited. Excessive layers of dust, that is layers greater than 5mm thick, are not covered by the presently available Standards, and where, for example in dust handling equipment, there may be thicknesses which exceed 5mm, specialist advice should be obtained.

Because the risk of ignition is from hot surfaces rather than directly from arcs and sparks, protection concepts for dust hazards concentrate on preventing the ingress of dust to internal parts which could present hot surfaces, and then establishing that the maximum temperature of the external surface will not, with a reasonable safety margin, exceed the ignition temperature of the hazard.

The normal methods of protection ('d', 'e', 'p', 'i' etc.) do not apply to dust hazards, and because the sealing (IP rating) of such methods of protection may not necessarily be adequate to prevent ingress of dust, it should not be assumed that any of these methods of protection are necessarily suitable. However, it is intended that there will be a guide to the use of BS 5501 electrical apparatus for use in dust hazards. At the time of publishing this book, the document is completing its Draft for Public Comment Stage, and after further review by the appropriate committees, will be published as a National British Standard. There are no definite plans at present to incorporate the proposed Standard into CENELEC work.

The main Standard which refers to electrical apparatus for dust hazards is BS 6467. This chapter is based on the guidance of this Standard, but also includes some reference to BS 5501 apparatus in anticipation of the publication of the new Standard before too long and on the assumption that there will not have been major changes in intent since the draft stage.

Certification of Apparatus

There is no Standard in the UK for the certification or approval of electrical apparatus for dust hazards, and the responsibility of any claim that apparatus is suitable for dust hazards rests with the manufacturer of the apparatus. The appropriate marking for such apparatus as stipulated in BS 6467 is considered later in this chapter.

Main Apparatus Design Criteria

Apparatus for use in dust hazard environments should be designed to minimize the accumulation of dust and electrostatic charges, and to facilitate cleaning.

Apparatus for use in Zone Z, should have a degree of protection which is dust tight in accordance with IP6X, and apparatus for use in Zone Y should have either dust tight or dust protected (in accordance with IP5X) enclosures.

The maximum surface temperature of the enclosure under adverse conditions of voltage is determined, and 10K is added to this figure to establish a safety factor. The value is corrected for an ambient temperature of 40°C, or such other temperature if a higher ambient temperature has been specified.

It is recommended by BS 6467: Part 2, that enclosures should be used which do not have maximum temperatures in excess of two-thirds of the ignition temperature (in °C) of the combustible dust, or 75K less than the ignition temperature of a 5mm layer of dust, whichever is the lower. This safeguard is particularly aimed at providing a safety margin to take account of the possibility of processes and procedures changing which could lead to the presence of dusts other than those envisaged at the design stage. The ignition temperatures for a selection of dust hazards is given in Table 5.1.

DUST	IGNITION TEMPERATURE OF LAYER IN °C
NATURAL PRODUCTS	
Wood, Flour	340
Cork	300
Cellulose	410
Grain dust	300
Spray dried, skimmed milk	340
Starch	435
Tea	300
Sugar, Milk	460
ORGANIC PRODUCTS	
Activated charcoal	400
Lignite	225
Phenolic resin	>450
Synthetic rubber	240
Methyl cellulose	320
Calcium stearate	>570
METALS	
Aluminium	>450
Bronze powder	260
Sponge iron	390
Zinc dust	440
INORGANIC PRODUCTS	
Lampblack	570
Molybdenum disulphide	320
Toner	>450

Table 5.1   *Ignition Temperatures of Layer of Dust* [4]

NOTE
The values are typical only. Precise values will depend on particle size, distribution, moisture content etc.

Enclosures shall be able to withstand an impact test of 7Nm. (Reduced to 4Nm for light transmitting parts with a guard, and 2Nm for light transmitting parts without a guard.) (The test procedure for impact tests is described in Appendix 9.)

Flanged joints should have a dust path width of at least 5mm, and be constructed such that when made up, the flange faces are substantially brought together over the joint face. Where gaskets are used on a flange joint, the dust path may be reduced to a minimum of 3mm. The gasket should be permanently attached to one or other of the faces to prevent its loss, or incorrect fastening.

It will be appreciated from the foregoing, that the resulting enclosure will not be dissimilar from a conventional flameproof 'd' design. Indeed, standard flameproof enclosures are often used for dust hazards, but the user should be aware that unless the enclosure includes a sealing gasket [5] it may not give the required IP rating protection.

It is the intention of the British Standard Guide to the use of BS 5501 electrical apparatus for use in the presence of combustible dusts to make reference to the methods of protection suitable for use in gas and vapour hazards. The following table indicates the likely recommendations.

Code letter for method of protection	Zone of use		Notes
	Z	Y	
o	OK	OK	1
q	OK	OK	2
p	OK	OK	
d	OK	OK	2, 3
e	OK	OK	2
ia/ib	OK	OK	2, 3
m	OK	OK	
N		OK	1, 3

Table 5.2   *Use of Normal Methods of Protection in Dust Hazards*

NOTES

1 Degree of protection to be at least IP5X. If the dust is electrically conducting, degree of protection to be IP6X.

2 Degree of protection to be IP6X if dust is electrically conducting or the classification is Zone Z.

3 Degree of protection to be at least IP5X

In general, the recommendations for installation, inspection and maintenance, are similar to those for the appropriate method of protection with gas or vapour hazards. Reference should be made to the appropriate chapter.

Marking

British Standard BS 6467 recommends the following information should be included in the marking of electrical apparatus suitable for use in dust hazard environments.

- The name or registered trade mark of the manufacturer.

- The manufacturer's type identification.

- The number and date of the British Standard (eg BS 6467: Part 1: 1985)

- The symbol (eg IP65) and reference of the appropriate British Standard (eg BS 5940) indicating the degree of protection.

- The maximum surface temperature.

- The maximum ambient temperature if other than 40°C.

- The minimum ambient temperature for use and storage if other than -20°C.

- Any applicable constraints on mounting position and orientation.

and, in addition, if the apparatus has been assessed by a recognized Testing Authority,

- The type approval certificate number

- The registered name or mark of the Testing Authority

- The letter 'X' after the type approval certificate number if there are special conditions of use which apply.

plus, if appropriate

- A warning 'Do not separate when energized' (applies to some plugs and sockets)

- A warning 'Do not open when energized' (applies to luminaries and certain switchgear)

Notes and References

1 British Standard Guide to the use of BS 5501 Electrical apparatus in the presence of combustible dust. This proposed Standard is currently at the 'draft for public comment' stage. It is anticipated that it will be produced as a National Standard during 1991.

2 May be referred to as zone 10 and 11 in Germany, or Division 1 and 2 in the United States.

3 Definition from BS 6467.

4 For more complete details refer to BS 6467: Part 2.

5 See Chapter 9 on the use of gaskets with flameproof apparatus.

6 Explosives Manufacture and Processing

Special regulations exist regarding the manufacture of explosives. For example, in the UK, Health & Safety document X1/6015/4/8A defines hazard categories A, B and C, and defines subdivisions as indicated in the following table.

CATEGORY	A	B	C
TYPE OF RISK	GASES AND VAPOURS	DUSTS	NOT DEFINED: EXPLOSIVE NORMALLY COVERED, AND THEREFORE VERY REDUCED RISK
SUBDIVISIONS	NORMAL ZONING AS FOR GASES AND VAPOURS (ZONES 0, 1, 2)	NORMAL ZONING AS FOR DUSTS (ZONES Z, Y)	

Section 2

NOTE

Although, for each method of protection discussed in this section, the main design and construction criteria are given, designers and applicants for the certification of apparatus should always consult the latest edition of the applicable Standard.

These chapters are not intended to be regarded as a definitive list of requirements for any method of protection.

Unless stated in the text, the information given in this section is based on the requirements of the European Standards EN50 014 etc.

CHAPTER 6

Non-Incendive and N-type Protection

Non Incendive and N-type Protection

SUMMARY	
CODE	Ex n Ex N[1]
STANDARDS	IEC 79-15 BS 4683: Part 3 BS 6941
UK CODE OF PRACTICE	BS 5345: Part 7
ZONE OF USE	Zone 2 only
LIVE MAINTENANCE	No

Introduction

N-type protection was initially seen as a way of using 'good quality industrial grade apparatus' which was non-sparking in normal operation, in zone 2, the least severe of the hazardous zone classifications. Because of this emphasis, it was not foreseen that certification for N-type protection would be common, and the original Standard for N-type apparatus (BS 4683: Part 3) does not contain sufficient information to fully define test and evaluation procedures.

It soon became apparent, however, that it was often difficult to establish just what was, and was not, sparking in normal operation. For example, is a screwed terminal a non-sparking device? What happens if it comes loose? Is it necessary to consider the loosening of a terminal, or is it exempt if it can be shown to be proof against loosening by vibration for example?

Since different manufacturers, wanting to sell apparatus to the hazardous area industry have very different ideas, ranging from excellent engineering standards to very poor practice, the end user began to ask for N-type apparatus to be certified, and not to rely on self certification by the manufacturer.

BASEEFA found it necessary to produce their own internal specification memorandum to clarify the contents of BS 4683: Part 3, so that N-type apparatus could be evaluated and tested. Then in 1988, a new British Standard; BS 6941 was issued to bring the UK more into line with the IEC 79-15 document. It should be understood that even BS 6941 is not necessarily intended as a certification Standard. The Standard recognises that certification by a third party may be desirable, and states that where such certification has been obtained, the mark of the certifying authority should be included in the label details. (See section of this chapter on marking).

N-type protection is not, at the time of writing, a CENELEC recognised method of protection. Thus there is no EEx N, and N-type apparatus will not display the Distinctive Community Mark. It appears that some EEC countries are now suggesting that only EEC certified apparatus should be used for new projects, and thus it may be that N-type protection will become less popular in coming years. This would be a pity, since N-type protection has considerable economic advantages over Ex d and Ex e for some types of apparatus, and there is no evidence to suggest that it does not provide an adequate level of safety for zone 2 applications.

In the USA, there has recently been a strong upsurge in the popularity of non-incendive protection. Although the USA technique is, in some respects more similar to increased safety 'e', it is only allowed in North America's Division 2, and thus seems better compared to Ex N than any other type of protection.

Main Constructional Requirements (BS 6941)

Ingress Protection

Apparatus enclosures for Ex N normally provide a degree of protection of IP54. In some cases, where live parts are insulated, IP44 may be acceptable. Where the ingress protection is less than these requirements, an 'X' will be included in the apparatus marking, and the certificate will specify special conditions of use.

Gaskets necessary to attain the required IP level must be fixed to one face of the joint which they protect, in order to prevent accidental assembly without the gasket being present.

Mechanical Strength

Enclosures are required to afford a reasonable degree of protection against impact. The severity of the impact test is determined by the type of apparatus/part under consideration. A distinction is also made between parts with a normal and low risk of mechanical damage, with the latter category subject to a less severe test. Tests are made by dropping a mass of M kg through a height of h metres onto the test sample or face of the apparatus. Table 6.1 shows the impact levels required by BS 6941 and includes the mass/height relationship required to achieve such levels. [2]

Additionally, portable apparatus is required to withstand a drop test of height 1m onto a flat surface. The test is repeated 4 times. Apparatus is considered satisfactory if, at the conclusion of these tests, the required level of ingress protection is not degraded.

TYPE OF PART OF APPARATUS	IMPACT ENERGY E (JOULES) RISK OF MECHANICAL DAMAGE =		MASS M	HEIGHT h
	NORMAL	REDUCED	(kg)	(m)
Light transmitting parts with guard (tested without guard)	1.0	0.5	0.25 0.25	0.2 0.4
Light transmitting parts without guard	2.0	1.0	0.25 0.25	0.4 0.8
Enclosures or parts of enclosures other than above	3.5	2.0	0.25 1.0	0.8 0.35

Table 6.1 *Mechanical Impact Tests*

Main Electrical Requirements

Since the main precaution of Ex N is to ensure there is no arcing or sparking in normal operation, the Standard sets down a number of requirements to ensure that this state is achieved. Many of the requirements fall into the realm of good engineering practice, for example the requirement that internal wiring shall be prevented from risk of contact with, and chafing against, sharp metallic parts by mechanical protection or clamping. It is also a requirement that contact pressure at terminals shall not be dependent on the pressure of insulating materials. Some requirements are less specific. For example, plugs and sockets are deemed to be separable under normal conditions, thus becoming sparking parts, unless the separating force required to part them exceeds 15N, or they are prevented from separating or loosening by mechanical means.

Terminals for external connection must be such that conductors can be readily connected to the terminals, and can be clamped in such a manner that they are gripped and secured against twisting and loosening, and that contact pressure is maintained.

Where plugs and sockets are used for external connection, they must be mechanically interlocked such that separation cannot occur whilst the contacts are energized, and so that they cannot be energized whilst the mating parts are separated. If these requirements cannot be met, then they must be secured together by fasteners or locking devices, and the apparatus must bear a label stating 'Do not connect or disconnect whilst energized'.

Insulation

Unless a circuit is intended to be directly connected to the case or frame of the apparatus, the construction must be such that an insulation test of voltage as indicated in Table 6.2 can be withstood for one minute without breakdown.

APPARATUS WITH WORKING VOLTAGE AND INTERNAL VOLTAGES NOT EXCEEDING 90V PEAK	OTHER APPARATUS
500V rms (+5% -0%)	(1000 +2U) V rms OR 1500V rms, whichever is greater. (Tol: +5% -0)
Where U is the nominal supply voltage or maximum internal voltage, whichever is greater.	

Table 6.2. *Insulation Test Voltages to Frame of Apparatus*

In addition to the above insulation requirements, any live conducting parts of different potential must have at least the creepage and clearance distances stated in Table 6.3 for them to be considered as not subject to arcing or sparking. Where movable parts are involved, the requirements must be applied to the worst case position.

NOMINAL SUPPLY VOLTAGE OR NOMINAL VALUE OF VOLTAGE BETWEEN CONDUCTIVE PARTS		MINIMUM CLEARANCES AND SEPARATIONS			MINIMUM CREEPAGE DISTANCE			
					MINIMUM VALUE OF COMPARATIVE TRACKING INDEX (CTI) [3]			
a.c V	d.c V	ENCAPSULATED mm	SEALED mm	IN AIR mm	500 mm	250 mm	175 mm	125 mm
12	15	0.13	0.3	0.4	1.0	1.0	1.0	1.0
30	36	0.26	0.3	0.8	1.0	1.0	1.0	1.0
60	75	0.43	0.43	1.3	1.3	1.3	1.3	1.3
130	160	0.66	1.0	2.0	1.4	1.7	2.0	2.5
250	300	0.66	1.7	2.0	2.3	2.8	3.4	4.0
380	500	0.73	2.6	2.8	3.7	4.3	5.1	6.7
500	600	0.9	3.0	3.4	4.4	5.1	6.0	7.1
600	900	1.1	4.4	5.0	6.5	7.5	9.0	11.0
1000	1200	1.7	5.8	6.8	8.6	10.0	12.0	14.0
3000	-	-	-	23.0	28.0	35.0	42.0	60.0
6000	-	-	-	45.0	55.0	70.0	85.0	-
10000	-	-	-	75.0	80.0	100.0	-	-

Table 6.3. *Minimum Clearances, Separations and Creepage Distances*

Specific Types of Apparatus and Circuits

Rotating Machines

N-type protection is commonly used for rotating machines. In the UK, a separate British Standard, BS 5000: Part 16 refers to rotating machines for Ex N, but this is not a certification Standard. If such apparatus is certified, it is certified to BS 6941.

Again, N-type protection is concerned with safety under normal operation. Thus as far as the temperature (and hence T-Class) of rotating machines is concerned, starting conditions are not included, unless the machine is designed to operate under duty cycle conditions when the maximum surface temperature determined by the temperature class of the machine shall not be exceeded for the whole of the duty cycle period, including starting or continuous stall, if this is part of the duty cycle.

Rotating machines and other devices with rotating parts should be constructed with mechanical clearances such that frictional sparking and hot surfaces are not likely to occur in normal operation. BS 6941 does not state what clearances are considered satisfactory for this aspect.

Fuses

Providing they are non-rewirable, non-indicating cartridge types, operating within their rating, fuse links are deemed to be non-sparking devices for N-type purposes. Thus the rupture of a fuse link is not considered to be normal operation. However, the temperature on the external surface of the cartridge fuse at the fuse's rated current must be taken into consideration in determination of the appropriate T-class for the apparatus.

The mounting apparatus for the fuse must be considered. When voltages of less than 60V are concerned, fuses may be soldered in place or retained in non-sparking spring holders or mounted in non-sparking enclosed holders. Above 60V, only the latter method of mounting is permissible.

Luminaries

Lighting fittings are a popular application for Ex N. Virtually all types of lamps are permitted, filament and tubular fluorescent, but low pressure sodium vapour lamps and other lamps containing free metallic sodium are excluded.

The lamp rating must be included in the label information, and a warning notice to prevent lamps being changed whilst energized.

Where a lamp requires a starter device, and where such device includes contacts, the contacts must be enclosed inside an hermetically sealed enclosure.

Instrumentation

Instrumentation, measurement and control applications frequently use Ex N protection, and may meet the requirements of BS 6941 even if certain aspects of the Standard are not directly complied with. For example, providing the voltage in the apparatus being considered does not exceed 60V ac or 75V dc, then relaxations on creepage and clearance distances may be permitted provided that the enclosure affords at least IP54 protection, (specific location apparatus, where the location provides such protection may be exonerated from this requirement) and that provision is made to ensure that the rated voltage is not exceeded in service.

Arcing and Sparking Parts and Hot Surfaces

Apparatus which does produce arcs or sparks, or hot surfaces in normal operation may be considered within the scope of Ex N, providing it can satisfy one of the following:

    an enclosed break device*
    non-incendive component*[4]

hermetically sealed device
sealed device
encapsulated device
energy limited apparatus and circuits
restricted breathing enclosure (hot surfaces only)

*maximum internal free volume allowed $= 20 \text{ cm}^3$

Perhaps the most normal approach for sparking contacts, especially with instrumentation, is the use of an energy limiting circuit. The criteria are the same as for the technique of intrinsic safety (see Chapter 13) except that no fault criteria need be considered, and no factor of safety is required for assessment against the ignition energy curves.

Non-incendive parts are defined as those parts with contacts for making or breaking a potentially incendive circuit, where either the contacting mechanism or the enclosure in which the contacts are housed is so constructed that ignition of the appropriate flammable gas or vapour is prevented under specified operating conditions. Since a non-incendive part must be assessed for a particular circuit, it is not possible to separately assess such a device as complying with the Standard, and thus it is not possible to produce such a device for general use.

Marking and Documentation

Apparatus should normally be marked with the following information:

- Manufacturer's Identification

- Manufacturer's Type Number or Reference

- The reference BS 6941:1988

- The symbol Ex N

- The Gas Group (or gas group and sub group if apparatus contains normally sparking parts)

- The temperature class

- The ambient temperature range if different from $-20°C$ to $+40°C$

- The symbol X if there are special conditions of use or installation

- Details of lamp rating etc. as appropriate

- Warning notice to prevent lamps being changed whilst energized, if appropriate.

Plus, if the apparatus has been certified,

- The name or symbol of the certifying authority

- The certificate number

Where apparatus has been certified by a third party, certification details will normally define any conditions of use. However, it is a requirement of BS 6941 that if so requested by the purchaser, the manufacturer shall provide documentation which includes:

A claim of compliance with the Standard or a certificate issued by an appropriate organisation.

Identification of the manufacturer.

Type identification.

Apparatus description.

Markings and any other information required by the appropriate parts of the Standard.

Any conditions etc. imposed by an 'X' in the label.

Notes and References

1 Although both BS 4683: Part 3 and BS 6941 refer to method of protection Ex N, BASEEFA have in the past used the code Ex n on some certificates (and hence lables) of apparatus Certified to BS 4683: Part 3. All certification to BS 6941 will, however, use the code Ex N.

2 The test apparatus required for the values of impact level is described in Appendix 9.

3 British Standard BS 5901 gives full details for the method of determining the comparative and proof tracking indices for solid insulating materials under moist conditions.

The CTI of a material is a measurement of how easily the surface of the material (insulator) will be contaminated by carbon deposits resulting from the conduction of electricity under damp conditions. If the CTI value is low, this means that a permanent track may be formed after drying if there has been conduction whilst the surface is wet. The higher the CTI value, the less likely is a permanent track to be formed on the surface.

Although it used to be quite difficult to obtain, for example, printed circuit board material with a CTI value in excess of 300, many modern materials used for such purposes meet a CTI of 300 without difficulty.

4 This use and definition of non-incendive should not be confused with its meaning in North America.

CHAPTER 7

Powder Filling

Powder Filling

SUMMARY	
CODE	EEx q
STANDARD	EN 50 017 BS 5501: Part 4
CODE OF PRACTICE	BS 5345: Part 9 (Not yet Published)
ZONES OF USE	1 & 2 (Zone 2 only UK)
LIVE MAINTENANCE	No
TYPICAL USES	Power supplies and similar apparatus

Although powder filling is perhaps not an especially popular protection concept, it nevertheless has a useful part to play. The main requirement is that the apparatus to be protected should be surrounded with a sand or powder filling medium such that any sparking or hot surfaces which exist under the filling medium cannot ignite a surrounding atmosphere.

The code letter 'q' stands for quartz, and the filling medium used is often small glass or silica beads. It has been found that the constructional requirements of the Standard EN50 017 are suitable for all gases, and in fact the potentially explosive atmosphere has little influence on the design characteristics. Thus, protection concept 'q' is normally certified for all gas groups.

The main requirements for sand/powder filling protection can be considered by referring to Figure 7.1.

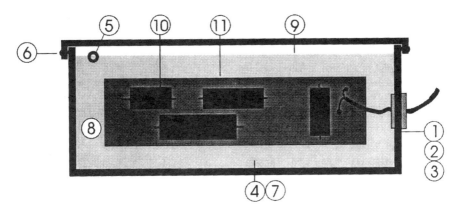

Figure 7.1   *Principal Requirements for Powder Filling*
(Apparatus without protective screen)

KEY TO FIGURE 7.1

1   Enclosure: normally metal

2   Enclosure must withstand pressure test of 0.5 bar overpressure
for 1 minute without distortion exceeding 0.5 mm in any
direction.

3   Enclosure to meet IP54 in normal service condition.

4   Filling medium: filling shall be carried out so that there are no
voids.

5   Viewing window to detect presence of voids.

6   Apparatus for Group I to have cover fixings complying with
'special fasteners' (see clause 8.2 of EN 50 014).

7   Filling medium normally quartz granules of size 250 μm to
1.6 mm. Filling material to contain less than 0.1% by weight of
water. Medium to be shaken down so as to avoid the formation
of voids.

8 Minimum distance from live parts to inner surface of enclosure to comply with Table 7.1.

9 Free surface of filling medium.

10 Minimum safe height (h_0) - not less than 30 mm for voltages of 1500V and less, and not less than 50 mm above 1500V.

11 Apparatus located in enclosure in such a way that in the operating position, it is entirely covered with a layer of filling material whose thickness above electrical parts, whether bare or insulated, must be at least equal to the minimum safe height h_0. (For electrical apparatus which can operate in more than one position, additional requirements apply.)

Rated Voltage V rms		Col A — Minimum distance between bare live parts and earth or between an insulated winding and the wall of the enclosure. mm	Col B — Minimum distance between a bare live part and the nearest point of the wall of the enclosure mm
	< 250	10	15
> 250	< 660	15	20
> 660	<1100	20	30
>1100	<3300	30	40
>3300	<6600	40	50

NOTE: For electrical apparatus with voltage not exceeding 500V, which is factory sealed in such a way that dismantling will destroy the apparatus, the value in Col A may be reduced to 4 mm, and the value of Col B reduced to 5 mm.

Table 7.1 *Minimum Distances Through Filling Medium*

Minimum Distances Through Filling Medium

Use of screens

In order to reduce the minimum safe height (h_0), a perforated metal screen may be located within the filling medium. The screen must be electrically bonded to the apparatus enclosure (unless the enclosure is of a non-metallic material). The screen must be perforated over the whole of its surface, with holes of diameter 8 to 10 mm, and on centres 50 to 70 mm apart. (This requirement may be reduced for apparatus with a total volume of less than 25 cm^3.)

The screen, together with its fixing mechanism shall be constructed in such a manner that it can withstand a force of 'P' Newtons at the centre of the screen, without causing a deflection of more than 1.2 mm.

Where:

P = 0.25 times the arcing current I_a [1]

(The arcing current is defined as the average value of rms current in an arc caused by a short circuit during its development within the filling medium of the apparatus.)

The position of the screen, relative to the enclosure walls and the apparatus, is defined as detailed in Figure 7.2.

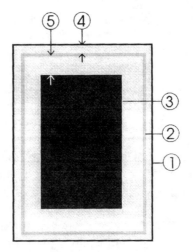

Figure 7.2 *Requirements for Metallic Screens in 'q' Enclosures*

KEY TO FIGURE 7.2

1   Enclosure

2   Screen (should completely surround electrical apparatus, unless the apparatus is only to be used mounted in a defined orientation)

3   Electrical apparatus

4   'Reserve layer'. Designed to fill up any voids which may occur. The minimum dimension is determined as at 20% of the protection height (h_{ea}) and at least 10 mm.

5   'Protection height'. (h_{ea}). Calculated from the equation:

$$h_{ea} = 0.31 \sqrt[3]{I_a^2 . t} \qquad \text{Units: cm}$$

Where:

I_a is the arcing current in amps

t is the duration of the arc in seconds

In all circumstances, h_{ea} must be at least 20 mm.

Marking and Information to be Provided

Apparatus should be marked as follows to conform to the general requirements of EN 50 014.

The name of the manufacturer or his registered trade mark

The manufacturer's type identification

The symbol EEx q

The symbol for the gas group (IIA, IIB, IIC)

The temperature class (T1-T6)

The indication of the testing station

The certificate number

and in addition be marked with:

The maximum permissible arcing current I_a and its duration $t^{(2)}$ as determined by the tests of the certification authority, unless, for small apparatus, an internal fuse is incorporated.

It is a requirement of the Standard that apparatus employing protection concept 'q' shall be supplied with an instruction sheet indicating the values of I_a and t, together with any variation to I_a if there are (for example by the use of a different circuit breaker) possible variations in time t.

The instruction sheet must also give details of the filling medium and topping-up procedure, if such is permitted by the certification conditions.

Notes and References

1   The short circuit arcing current (I_a) can usually be established
    for apparatus operating up to 6 kV from the relationship

$$I_{cc}/I_a \text{ is approximately} = 1.3$$

   where Icc is the prospective short circuit current of the input
   supply.

2   If the apparatus is protected by a circuit breaker, then the time
    't' is the total breaking time of this device.

CHAPTER 8

Oil Immersion

Oil Immersion

SUMMARY	
CODE	EEx o
STANDARDS	EN 50 015 BS 5501 Part 2
CODE OF PRACTICE	BS 5345: Part 9 Not yet published
ZONES OF USE	1 & 2 (Zone 2 only in UK)
LIVE MAINTENANCE	No
TYPICAL USES	Transformers, switchgear (see text), distributed control systems

Oil immersion, although established as a method of protection for many years, has never been especially popular, and indeed there may have been some confusion over the way this method of protection should be used. Traditionally, oil immersion has been used for heavy current apparatus, especially transformers and switchgear. In these applications, particularly with switchgear, the oil is actually quenching the electrical arc or spark. This is not a particularly desirable state, since mineral insulating oil, which the present European Standard EN 50 015 requires, breaks down under electrical arcing and produces hydrogen and acetylene! Although the quantities of these products is very small in relation to any external atmosphere, there is understandable nervousness with the possibility of the electrical apparatus actually producing a hazardous area.[1]

Another possible reason why oil immersion has never been popular is that if there are indeed sparking or arcing parts under the oil, then there only needs to be one failure, for example a failure which allows the oil to drain away, and the apparatus is left in a potentially ignition capable state. For this reason, the UK has only allowed oil immersion as a method of protection for zone 2, where it is considered that safety under normal operation is a sufficient safeguard. (Compare, for example with N-type protection.) Most other European countries, however, have allowed oil filling in both zones 1 and 2.[2] Until there is some CENELEC Standard, presumably based on the IEC 79-10[3] document which allows hazardous area zoning techniques to be common to all European countries, and thus allows each method of protection to be used in the same zones regardless of the country of installation, this difference is likely to continue.

There is, however, little disagreement that if oil immersion is used for apparatus which is not normally arcing, then the technique is suitable for both zones 1 and 2. Since this distinction is not currently recognised by the European Standard, the UK zone 2 only approach persists.

It has recently become necessary to review the CENELEC oil immersion Standard, since it is not a very comprehensive document, and does not include all the information needed for design and certification testing. A working party from the BSI committee GEL 114/3 has produced a draft Edition 2 of EN 50 015, and it seems likely that this document will be adopted, possibly with some minor amendments, as the new Standard.

In recent years there has been significant interest in the possibility of using oil immersion for lower power apparatus, and one company has successfully achieved a certified EEx o distributed control system, aimed particularly at the offshore industry. Oil immersion has a number of significant advantages for such uses. In particular:

− The presence of the oil greatly assists environmental protection.

− The ability of the oil to circulate by convection around the electrical apparatus submerged in the oil tends to disperse any hot spots. This leads to significantly improved reliability for the apparatus.

The new draft Standard recognises this possible advance in the use of oil immersion, and, for the first time, specifically excludes the technique from being used where there are normally arcing and sparking parts under the oil. The revised Standard also makes it possible to use a filling medium other than mineral insulating oil, thus opening the way for modern silicone fluids and vegetable oils to be used.[4]

In view of these likely changes emerging as a revised Standard within the next few years, and the fact that the existing Standard is so rarely used, the technical data in this chapter will concentrate on the draft revision of the Standard EN 50 015.

Main Requirements

Electrical apparatus complying with this method of protection must be submerged in the filling fluid to a depth of at least 25mm. (This requirement excludes interconnecting conductors - which must comply with the creepage and clearance requirements of increased safety[5] or be part of an intrinsically safe circuit. Also, apparatus other than conductors is allowed above the filling fluid if it is intrinsically safe.)

The normal mechanical strength requirements specified for all methods of protection under EN 50 014 (BS 5501 Part 1) apply to this protection method.

It must be possible to determine the level of the filling fluid under normal operation. This may be achieved by a sight glass or other similar means. Dipsticks are also permitted in certain circumstances.

If a draining device is provided, it shall comply with the requirements of 'special fasteners' described in EN 50 014. This in essence means that the likelihood of such a device being opened inadvertently, or by unauthorised personnel, is minimised by the need for special tools etc.

Both sealed and non-sealed enclosure designs are permitted.

The temperature of the filling fluid must not exceed a safe limit for its protection and electrical properties, for example must not exceed 105°C for Class II mineral insulating oils and 115°C for Class I Oils. (See IEC 296)

Sealed Enclosures

See Figure 8.1

If the enclosure is sealed, then it must be provided with a pressure relief valve. The outlet of the pressure relief valve must provide a degree of protection of at least IP23. The purpose of the pressure relief valve is to prevent, especially under fault conditions, a dangerous build up of pressure which could damage the enclosure. The pressure relief valve must be set to at least 1.1 times the maximum pressure which can be present in the enclosure under normal conditions. (This pressure will normally be the pressure exhibited when the fluid temperature is raised above the filling temperature, as a result of the expansion coefficient of the filling fluid, which may be about .0007.) In addition, the enclosure must withstand a type test and a routine production test for overpressure at 1.5 times the setting of the pressure relief valve for 60 seconds, without distortion or adverse affect on the IP66 requirement of the enclosure.

The enclosure must also pass a type and routine test for under pressure. The enclosure is evacuated to a pressure equal to a reduction in filling fluid level from the maximum to minimum permitted level. The pressure reducing apparatus is then isolated, and the enclosure left for 24 hours. At the end of the test there shall not be more than a 5% change (increase) in pressure. Temperature conditions must remain constant throughout the test.

Sealed enclosures must not use a dipstick as the filling level indicating device.

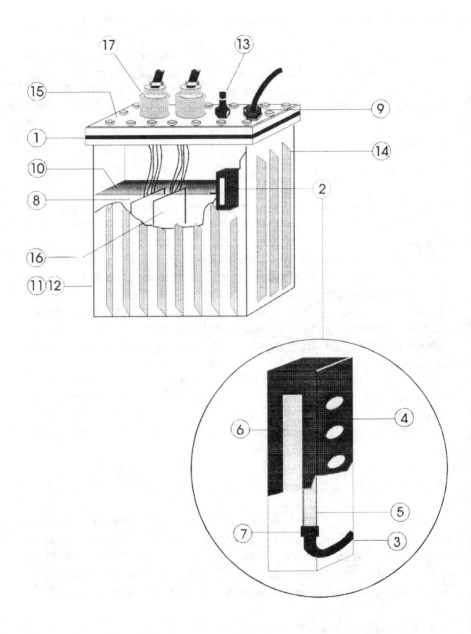

Figure 8.1 *Typical Oil Immersed Apparatus (Sealed Design)*

KEY TO FIGURE 8.1

1 Sealing gasket
2 Protection fluid sight glass
3 Pipe welded/brazed to enclosure
4 Sight glass cover welded/brazed or fixed with special fasteners
5 Sight glass
6 Width of window in sight glass cover such that impact test (see Appendix 9) of 7 Nm does not damage sight glass
7 Glass - metal join
8 Minimum depth of fluid of 25 mm over protected apparatus
9 Cable gland to withstand pressure tests and impact test
10 Protection fluid: mineral insulating oil etc.
11 Apparatus to withstand 7 Nm impact test
12 Apparatus to withstand over and under pressure test
13 Pressure relief device set at 1.1 times maximum normal operating pressure
14 Cooling fins (if required)
15 Cover fixing with special fasteners
16 Electronics to be protected
17 External connectors, for example for intrinsically safe circuits. Socket and its mounting must be such that pressure test requirements can be met.

Non-Sealed Enclosures

Apparatus which is not sealed must be constructed so that any gas or vapour which may evolve from the filling fluid in normal service can readily escape. Breathing devices and openings shall be provided with a suitable drying agent.[6] The outlet of the breathing device requires a degree of protection of at least IP23.

The filling level indicating device may be a dipstick, but any such dipstick must be secured in its measurement position during normal operation, and in this condition the IP66 rating shall be preserved.

An oil expansion facility is required, and the apparatus must be provided with a manually only re-setable protective device which automatically causes interruption of the supply current if there is an internal fault in the fluid filled enclosure which could cause evolution of gas from the filling fluid. This condition is likely to occur if very hot surfaces, or arcs and sparks become present under fault conditions.

Filling fluid requirements

The filling fluid must be either mineral insulating oil to IEC 296,[7] or an alternative fluid which complies (meets or exceeds) those requirements. Typical characteristics are shown in Table 8.1

Requirement of EN 50 015		Typical Mineral Insulating Oil to IEC 296	Typical value for silicone fluid (Dow Corning 561)	Typical value for Galden fluid DO2	Typical value for Midel (f)
Flash point 140°C min	(a)	150	> 300	NONE	
Kinematic viscosity 100 cSt	(b)	16	50	1.8	100
Electrical breakdown strength					
27 kV minimum	(c)	40	50	50	
Volume resisitivity					
10 ohm cm minimum	(d)	10	10	10	
Pour point -30°C max	(e)	-45		-97	-50
NOTES					
a Determined in accordance with ISO 2719					
b Max value at 25°C, determined in accordance with ISO 3104					
c Determined in accordance with IEC 156					
d Determined in accordance with IEC 247					
e Determined in accordance with ISO 3016					
f Determined in accordance with IEC 588-2					

Table 8.1   *Requirements for Immersion Fluid and Data for Typical Filling Fluids*

Marking

All the normal requirements for apparatus marking[8] as detailed in EN50 014 (BS5501: Part 1) apply:

–   The name of the manufacturer or his registered mark

− The manufacturer's type identification

− The symbol EEx o

− The symbol for the Gas Group of the electrical apparatus:
    I for mines susceptible to firedamp
    II or IIA or IIB or IIC for other applications

− The temperature class (T1 to T6) or the maximum surface temperature in °C, or both. If the maximum surface temperature for Group II apparatus exceeds 450°C, then only the temperature shall be marked, and not the T-Class.

− A serial number, except for small components, bushings etc.

− The indication of the testing station

− The certificate number, commencing with the year of certification.

− The letter X at the end of the certificate number if there are special conditions for safe use.

and, in addition, for fluid filled apparatus type 'o':

− An indication of the filling fluid to be used.

− The pressure relief valve setting, where a pressure relief valve is used.

Notes and References

1 Oil filling is not, of course, the only electrical apparatus which produces a hazard. Lead acid accumulators evolve hydrogen under charging conditions, and the release here is likely to be very much more than from the occasional arc or spark from oil filled apparatus.

2 Belgium, at the time of writing also restricts oil immersion to zone 2 use only.

3 IEC 79-10 Electrical Apparatus for Explosive Gas Atmospheres, Part 10: Classification of Hazardous Areas.

4 The technique of oil immersion using mineral insulating oil is not acceptable for Group I apparatus.

5 For details of 'increased safety' see Chapter 12. The creepage and clearance requirements appear in Tables 12.6 and 12.7.

6 It should be noted that silica gel drying agents may need to be changed on a regular basis if exposed to prolonged moist conditions. However, although the drying agent may become saturated, it is doubtful if such saturation will result in harm or deterioration of the filling fluid, and the apparatus design should, in any case, minimise the possibility of moisture contaminating the filling fluid by careful positioning and protection of any breathing device.

7 IEC 296 = BS 148, Unused mineral insulating oils for transformers and switchgear.

8 For examples of marking, see Appendix 1.

CHAPTER 9

Flameproof and Explosionproof Protection

Flameproof and Explosionproof Protection

SUMMARY	
CODE	Ex d or EEx d
STANDARDS	BS 229 BS 4683 Part 2 EN 50 018 BS 5501: Part 5 BS 5000: Part 17
CODE OF PRACTICE	BS 5345: Part 3
ZONES OF USE	1 & 2
LIVE MAINTENANCE	No
TYPICAL USES	Rotating machines Lighting, Junction boxes instrumentation, start/ stop buttons

Flameproof protection, or 'explosionproof' as it is known in North America, is probably traditionally the most widely used method of protection for electrical apparatus in hazardous areas. Although there have been numerous Standards in the evolution of the technique, the principle remains unaltered.

The requirement is for the apparatus enclosure to be of sufficient strength and integrity to withstand an internal explosion of the hazard for which it is designed, without the explosion setting fire to the surrounding atmosphere.

Until relatively recently, flameproof enclosures were nearly always made from cast iron, but in recent years the use of aluminium (providing it contains less than 6% magnesium by weight[1]), phosphor bronze, stainless steel and other materials have become more popular.

Bearing in mind that to avoid ignition, it is necessary to remove or control either the source of ignition, the supporter of combustion, or the hazard, it will be appreciated that the technique of flameproof protection, which assumes that all three properties will coexist within the enclosure, is very dependent on good mechanical design. It should also be remembered that environmental effects and poor maintenance are more likely to be degrading to flameproof apparatus than to other methods of protection.

Although there is much certified apparatus still available to the older Standards BS 229 [2] and BS 4683: Part 2 [3], in this chapter, the requirements of the European Standard EN50 018 (British Standard BS 5501: Part 5) will be discussed. In essence, the differences between the Standards involve aspects which will only affect the designer of flameproof apparatus, rather than the end user.

Main Requirements

As stated in the introduction, the main aim of flameproof protection 'd' is to prevent any flame (fire or explosion) which exists within the item of flameproof apparatus from igniting a surrounding hazardous atmosphere.

Flameproof protection is allowed in zones 1 and 2, and thus it must be assumed that there may be a potentially flammable mixture of gas/air present inside the enclosure as a result of the enclosure having been opened, or due to any gaps in the flanges and openings of the enclosure. With flameproof protection, the inclusion of arcing and sparking parts, or of hot surfaces inside the apparatus is not excluded, and so the possibility exists that an explosion may occur inside the apparatus. If such an explosion does occur, then damage to the interior of the apparatus is possible. Such damage to the electrical apparatus does not matter from the hazardous area viewpoint, **providing that the fire/explosion does not ignite the surrounding hazardous atmosphere.** [4]

Considering the example of Figure 9.1, which represents some hypothetical flameproof apparatus enclosure whose only opening

Figure 9.1   *Simple Flameproof Enclosure*

is its lid, it can be seen that if an explosion takes place within the enclosure, the possibilities for ignition of the surrounding atmosphere are:

- − if the enclosure itself explodes

- − if the enclosure is damaged by the explosion; for example if the lid or cover is not firmly held in position

- − if the external surface becomes hot enough to ignite the surrounding hazard by exceeding the auto ignition temperature. [5]

In practice, the third possibility is low, because an explosion will be completed very quickly, and will not present a risk of elevating the external surface temperature significantly. Even if the explosion causes a fire within the enclosure, the absence of any appreciable supporter of combustion (air) within the enclosure (it will have been used up by the explosion) will prevent the fire being sustained.

The main considerations then are the first two possibilities. Firstly, consider that the enclosure has no lid, and no opening of any sort at all. To meet the requirements of flameproof protection, it would need to withstand the pressure and effects of an internal explosion without fracturing. This position is similar to the simple flameproof enclosure where the flange at the joint between the enclosure and its lid is perfectly flat and smooth, giving a 'sealed' enclosure by its metal-to-metal join. In this situation, the enclosure will withstand an internal explosion. Indeed, it is a requirement of flameproof protection that the enclosure will withstand 1.5 times the internal explosion pressure. Where the flameproof apparatus includes welded joints, every enclosure will be static pressure tested to 1.5 times the experimentally determined internal explosion pressure.

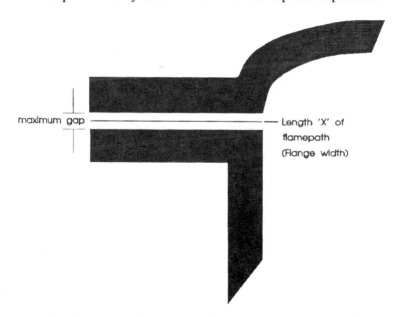

maximum gap — Length 'X' of flamepath (Flange width)

Figure 9.2 *Flange on Flameproof Joint*

Due to mechanical tolerances, and normal manufacturing techniques, it is likely that there will not, in fact, be a perfect join at the machined flange faces. There may be a small gap around the flange, at least at some points. The possibility of a gap is allowed for in the design and certification of the apparatus, and the presence of a long 'flamepath' (X) ensures that even with a gap, providing such gap does not exceed a maximum safe gap [6] for the enclosure size and the hazard concerned, then any internal flame will not ignite a surrounding hazardous atmosphere. (Figure 9.2)

It should be understood that this assurance is not absolute. The possibility of incendive flame being transmitted through the gap **of a correctly installed flameproof enclosure** has been determined experimentally, and it is believed [7] that the likelihood of such an occurrence is less than 1 in 10^5. This is considered to be an acceptable level of risk for zone 1 applications. [8]

Flameproof Joints

Since all flameproof apparatus will have some joint(s), the maximum permitted gap for joints of differing designs and for different Gas Groups is defined in the Standards. There are slight differences in the requirements, depending on the Standards used, but assuming the European Standard EN50 018, the following is a summary of the requirements.

GAS GROUP	MINIMUM JOINT WIDTH mm (Dim X in Fig 9.2)	MAXIMUM GAP FOR ENCLOSURES WITH VOLUME:			
		$\leq 100cm^3$	$>100cm^3 \leq 500cm^3$	$>500cm^3 \leq 2000cm^3$	$>2000cm^3$
I	6	0.3	-	-	-
	9.5	0.35	0.35	-	-
	12.5	0.4	0.4	0.4	0.4
	25	0.5	0.5	0.5	0.5
IIA	6	0.3	-	-	-
	9.5	0.3	0.3	-	-
	12.5	0.3	0.3	0.3	0.2
	25	0.4	0.4	0.4	0.4
IIB	6	0.2	-	-	-
	9.5	0.2	0.2	-	-
	12.5	0.2	0.2	0.2	0.15
	25	0.2	0.2	0.2	0.2
IIC	6	0.1	-	-	-
	9.5	0.1	0.1	-	-

Table 9.1   *Relationship Between Maximum Permitted Flange Gap and the Minimum Flange Width for Various Enclosures and Gas Groups*

NOTES FOR TABLE 9.1

1 Flanged joints are not permitted for gas group IIC apparatus intended for use where acetylene may be present.

2 Group IIC enclosures are not permitted to have flanged joints if the internal volume exceeds 500 cm^3.

3 There should be no intentional gap at a flange joint, but values shown are the maximum permitted if such a gap exists.

4 Surfaces of joints must have a machined surface such that the roughness does not exceed 6.3 μm.

Where the flange is interrupted by holes for fixing screws or studs, the maximum distance along the surface of the joint between the interior of the apparatus and the hole shall be at least that shown in Table 9.2. (See Figure 9.3)

MINIMUM DISTANCE TO HOLE y mm	JOINT WIDTH x mm
6	< 12.5
8	≥ 12.5 < 25
9	≥ 25

Table 9.2 *Requirements for Flanges Interrupted with Fixing Holes Associated with the Two Parts*

Threaded joints, including cable entry glands shall, for all gas groups, have a pitch of at least 0.7 mm, with medium or better quality of fit, shall have at least five threads engagement, with a depth of engagement of at least 5 mm for enclosures of internal volume not greater than 100 cm^3 and at least 8 mm for enclosures of greater volume.

Sealing Gaskets and Other Means of Environmental Protection

Since, as discussed, the design and manufacturing constraints of flameproof apparatus may mean that there are gaps at jointed parts, it follows that flameproof protection often affords less environmental

128

protection than is required for the installation. It is thus frequently necessary to improve the environmental properties of flameproof apparatus by various means - gaskets, grease, tape, etc.

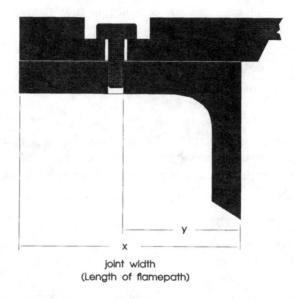

Joint width
(Length of flamepath)

Figure 9.3 *Flanges with Interruption of Holes*

Where gaskets are to be used, they must be provided for in the apparatus design, for example by the presence of a purpose placed groove for an 'O' ring etc. (Figures 9.4 and 9.5 give examples) In no instance shall the presence of a gasket or other compressible material be allowed to interfere with the intended dimension of flange width or gap. Bearing in mind that intentional gaps are not permitted in flange joints, this rules out the possibility of using any gasket with a flange design as shown in Figures 9.1 and 9.2. It will be seen from the examples of Figures 9.4 and 9.5, that where a design includes provision for a gasket, the minimum flange width (length of flamepath) and maximum gap dimension are still preserved in all situations.

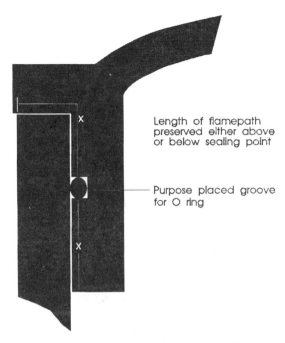

Length of flamepath preserved either above or below sealing point

Purpose placed groove for O ring

Figure 9.4   *Example of Flameproof Joint With Sealing 'O' Ring*

Although the use of grease and tape for improving the environmental properties is not addressed in the European Standard, it does receive attention in various Codes of Practice; for example the UK Code of practice, BS 5345: Part 3.

Grease should be of a type which does not set hard. The following types are acceptable:

Proprietary makes specially designed for flameproof apparatus, e.g. Hylomar sealant (BAS1050U)

Silicone grease*

Vaseline

* Note that silicone grease can adversely effect some gas detectors. Care should be taken, or the advice of the gas detector manufacturer sought before silicone grease is used in the vicinity of such applications.

130

Where tape is used around the outside of a completed flange joint
the requirements are as follows:

- In any circumstances, only one layer of tape should be used.
  A small overlap is permitted to enable the tape to be secured.

- Tape is not permitted for group IIC enclosures.

- For group IIB enclosures, the advice of the manufacturer or
  suitable expert should be sought.[9]

- For group IIA enclosures, tape may be used, without
  reference to the manufacturer.

- If the tape is removed for any reason, new tape should be
  applied on re-assembly.

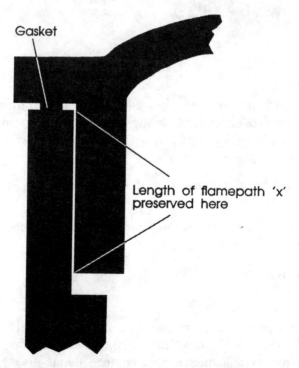

Figure 9.5 *Example of Flameproof Joint with Sealing Gasket*

Obstructions Outside Flameproof Apparatus

It has been found that solid obstructions which are not part of the apparatus, near to the edges of flanged joints or openings, may impair the efficient operation of a flameproof enclosure. Thus, care should be taken to prevent solid obstacles such as steel work, walls, pipes or other electrical equipment approaching closer than the following to any flameproof flange or opening.

Group	Minimum distance from flange to obstruction (mm)
IIC	40
IIB	30
IIA	10

Table 9.3   *Minimum Obstruction Distances to Flameproof Joints*

It should be noted that this restriction often precludes the popular practice of mounting small junction boxes and switchgear in the channel section of vertical 'H' section steel girders.

Cable Entries

Cable entries to flameproof apparatus are achieved by either 'direct' or 'indirect' means. Cable glands are an example of a direct cable entry, (Figure 9.6) and it should be realised that the cable gland must be of a certified type (normally a Component Certificate) [10] and be of the correct type (thread) for the gland entry hole in the apparatus. Cable entry holes will be provided by the manufacturer, or possibly a variety of gland plates will be available. The end user should not attempt to make additional cable entry holes unless he has the permission of the manufacturer and is competent to do so. Where possible and practicable, it is good advice to obtain apparatus with additional cable gland entry holes which can then be blocked using the appropriate blanking gland, available from the cable gland manufacturers.

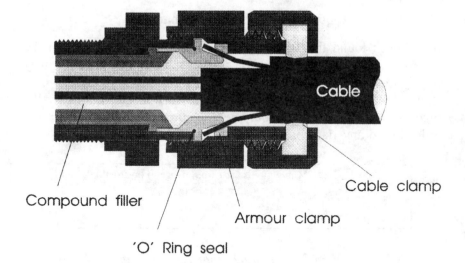

Compound filler

'O' Ring seal

Armour clamp

Cable

Cable clamp

Figure 9.6   *Example of flameproof Stopper Gland*

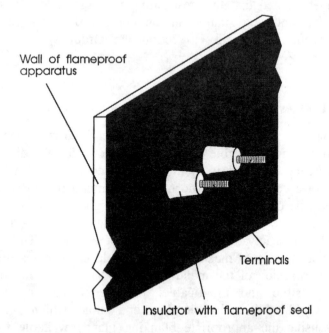

Wall of flameproof
apparatus

Terminals

Insulator with flameproof seal

Figure 9.7   *Basic Arrangement for Indirect Cable Terminations*

Sometimes solid conductors are brought out from a flameproof enclosure to a separate terminal compartment. In this instance, the terminal compartment usually employs the protection technique of increased safety 'e'. Such arrangements (shown for example in Figure 9.7) are known as indirect cable entries.

Where interconnecting cables to flameproof apparatus utilise conduit, it should be noted that it is essential to ensure that a sufficient seal is provided between the flameproof enclosure and the conduit. This usually involves the use of a suitable stopper box.

Failure to correctly seal the conduit could invalidate the security of the flameproof apparatus, or lead to pressure piling which may destroy the flameproof enclosure or allow flame transmission by rupturing the conduit.

The use of plastic, aluminium or flexible conduit is only normally permitted in zone 2.

Pressure Piling in Flameproof Apparatus

When interconnected compartments are present within flameproof apparatus, there is a danger of pressure piling. This is the effect caused by an explosion in one compartment increasing the pressure such that when flame propagates to a second compartment to cause a secondary explosion, that explosion, starting from an already increased pressure may damage the flameproof integrity of the apparatus.

Pressure piling can occur in sectionalised enclosures, between the end sections of a motor if the gap between the rotor and stator is too great; caused, for example by cleaning the rotor in a lathe, and in other similar situations. (Figure 9.8)

134

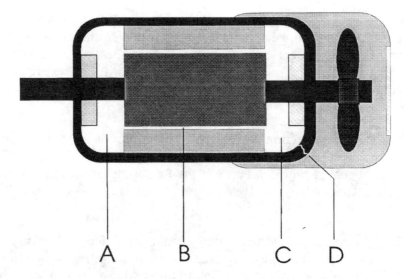

Figure 9.8 *Pressure Piling in Motors*

NOTE: An explosion in compartment 'A' pressure piles via the gap 'B' between the
stator and rotor, which has been increased above the certified maximum by
incorrect cleaning techniques. This causes a secondary explosion at 'C'
which, as an explosion under increased pressure, may crack the casing 'D'
and destroy the enclosure.

Rotating Machines to BS 5000: Part 17

Many motors for hazardous areas bear a label stating conformance
to BS 5000. This Standard is a general Standard for rotating
machines, with different parts of the Standard referring to specific
requirements. Part 17 relates to machines with flameproof
enclosure. This Standard refers closely to one of the main
construction Standards: BS 229, BS 4683: Part 2 or BS 5501: Part
5. It should, however, be noted that labelling indicating compliance
with BS 5000: Part 17, does not necessarily mean that the motor is
certified. If it is certified, the normal certification details will
appear on the label in addition to any claim to comply with BS
5000: Part 17.

Breathing and Draining Devices

It is quite common for flameproof apparatus to require breathing or draining devices for operational purposes. Such devices are allowed within the scope of the European Standard, but certain arrangements, for example the use of sintered discs is not covered. Where sintered discs are used, it is common for certification to make reference to special protection techniques, that is by a coding of Ex ds, to cover this aspect. This matter should be resolved when the second edition of EN 50 018 is published.

Maintenance of Flameproof Apparatus

It should by now be appreciated that flameproof protection is particularly vulnerable to deterioration of the enclosure from rust, corrosion by chemicals etc. It is of crucial importance to the integrity of this method of protection that regular inspections and maintenance are effected to ensure that the intended level of security from ignition is maintained. Chapter 17 discusses this aspect in more detail.

Marking and Labelling

Flameproof apparatus should include the following label information.

A. Apparatus to EN 50 018

Manufacturer's name
Type number/serial number
Mark of Certification Authority
Distinctive Community Mark (if appropriate)
Certificate number and name of Certifying Authority
The code EEx d
The code for the Gas Group (e.g. IIA, IIB, IIC)
The code for the temperature class (T1 to T6)
A warning indicating that the enclosure should not be opened whilst energized.

Plus, for motors etc.

A warning label concerning any live parts e.g. anti-condensation heaters or thermocouples which will still be energized after removal of the main power.

B.   Apparatus to BS 4683: Part 2

Manufacturer's name (or name of agent if other than the manufacturer is responsible to the National Testing Authority)
Manufacturer's type identification
The mark BS 4683: Part 2
The gas group
The Certificate number
Name or mark of the Certifying Authority
Temperature Class

Plus

Any other relevant information

Plus (optional)

The mark IEC 79-1 if the enclosure also meets the requirements of that Standard.

North American Explosionproof Apparatus

Although the main aspects noted in this chapter apply to 'explosionproof protection' in North America, there are some constructional differences, and the reader who requires specific information on apparatus design for the American market should consult the appropriate Standards and other documents.

Factory Mutual Research Corporation, Underwriters Laboratories and the Canadian Standards Association all publish Standards specific to explosionproof protection. The main Standards are listed at the end of this chapter. [11]

Notes and References

1 See EN50 014. It is a general requirement that enclosures for electrical apparatus in hazardous areas contain less than 6% magnesium by weight. This is because of the increased risk of incendive sparks with magnesium.

2 BS 229: Flameproof enclosure of electrical apparatus.

3 BS 4683: Part 2: The construction and testing of flameproof enclosures of electrical apparatus

4 It should be noted that although we refer to 'methods of protection for electrical apparatus', it is, of course, the hazardous area we are trying to protect; not the electrical apparatus itself.

5 For explanation of Auto Ignition Temperature, refer to Chapter 2.

6 This value is not necessarily the same as the maximum experimental safe gap, which is defined as the maximum gap of a joint of 25mm width which prevents any transmission of an explosion in 10 trials made under specified conditions. (See IEC 79-1A)

The MESG values are as follows:

GROUP	MESG (mm)
I	0.8
IIA	0.65
IIB	0.35

It will be appreciated that the values of maximum gap permitted are 2/3 of the MESG, to allow for a factor of safety of 1.5.

7 See Hicks, W.K. and Brown, K.J. Assessment of Explosion Probability for Intrinsically Safe Apparatus. IEE Conference, Electrical Safety in Hazardous Environments, 1971.

8 Using numerical values for zone hazard present time discussed in Chapter 2, a zone 1 hazard may exist at potentially flammable concentrations for between 10 and 1000 hours per year. Taking the logarithmic mid way point of 100 hours per year (that is, about 10% of the maximum time for zone 1) gives a statistical likelihood of ignition from correctly installed and maintained flameproof apparatus as $P_c = 10^{-5}$. This figure is useful to compare against statistical analysis for other methods of protection in other zones. For example, to achieve the same level of protection in zone 2 as is achieved for flameproof protection in zone 1, we need a probability of ignition to be no more than 10^{-4}, since the likelihood of the hazard being present is 10 times less. Similarly for zone 0, the value should be 10^{-6}, since the likelihood of the hazard being present is ten times greater.

9 In practice, few end users bother to consult the manufacturer but, if consulted, few manufacturers object!

10 See Chapter 4 for details of Component Certificates.

11 The following Standards are widely used for the design and construction of explosionproof apparatus in North America.

FM3165	Explosionproof protection
UL 698	Explosionproof industrial control equipment
UL 866	Explosion outlet boxes and fittings
CSA c22.2 No 30	Explosionproof apparatus

CHAPTER 10

Pressurization

Pressurization

SUMMARY	
CODE	EEx p
STANDARD	EN 50 016 BS 5501: Part 3 IEC 79-2 NFPA 496 (USA)
CODE OF PRACTICE	BS 5345: Part 5
ZONES OF USE	1 & 2
LIVE MAINTENANCE	No
TYPICAL USES	Control panels, VDU's analyzers.

Pressurization is a well established method of protection and is widely used in hazardous areas. Although its main use is for larger apparatus such as control panels, there is no reason why it should not be used for smaller applications. Although usually a custom built approach, possibly using a component certified control system, there are some standard 'off the shelf' applications such as pressurized cabinets for personal computers and VDU's.

The concept of pressurization is very simple; namely to keep an enclosure free from hazard by supplying it with a supply of clean air from outside the hazardous area, or protective gas. Because various safeguards are needed to ensure that the enclosure has been properly protected, the actual mechanics of the technique are actually a little more complex than might at first be thought.

In the USA, three categories of pressurization are recognised:

X purging   which takes the inside of the purged apparatus from a Division 1 to a non-hazardous area.

Y purging   which takes the inside of the purged apparatus from a Division 1 to a Division 2 area.

Z purging   which takes the inside of the purged apparatus from a Division 2 to a non-hazardous area.

The requirements for each category vary, for example Z purging does not require any safeguards of the purging system, whereas Y purging limits the apparatus within the purged apparatus to that which is non-sparking or non-incendive in normal operation. Type X purging is the most stringent category, since it allows ignition capable apparatus within the enclosure. For this category, the safeguards required include an electrical interlock against purge failure, and precautions against accidental opening of the enclosure whilst the equipment is live.

The detailed requirements are given in NFPA 496-1986, published by the American National Standards Institute.

In essence, there is little difference between the practice applied in the USA, and the advice of the UK code of practice for pressurization, BS 5345: Part 5, which recommends that pressurized apparatus which contains ignition capable parts, located in zone 2, and pressurized apparatus which contains no normally ignition capable parts, located in zone 1 should alarm on pressure failure, whereas pressurized apparatus containing normally ignition capable parts which is located in zone 1 should both alarm and de-energize if the pressure fails. Pressurized apparatus which contains no normally ignition capable parts, located in zone 2, presumably already (at least in essence, even if not certified) complies with the requirements for zone 2 apparatus without the need for the pressurization, and thus this option is not covered in terms of advice for alarms etc. in the Code.

Main Requirements (EN 50 016)

Type of protection 'p' is defined as a technique of applying a protective gas to an enclosure in order to prevent the formation of an explosive atmosphere inside the enclosure by maintaining an overpressure against the surrounding atmosphere, and where necessary by using dilution.

The main document relating to the design and certification principles of protection concept 'p' is the CENELEC Standard EN 50 016. This Standard is soon to appear in its second edition, and the changes between Edition 1 and 2 are quite significant. The second edition of the European Standard EN 50 016 is, at the time of publication awaiting the results of the public comment stage, and it seems likely that a number of alterations may still be made before it is finally published. However, there seems little doubt that the second edition will be a far more comprehensive document than the existing Edition 1, and this chapter has taken the new work as the basis of the explanations given. If the reader requires precise information, the latest edition of the Standard should be consulted.

There are three main ways in which the technique of Ex p can be achieved:

Static pressurization

Pressurization with leakage compensation

Pressurization with continuous flow [1]

In addition to the above, there are additional requirements if the pressurized apparatus contains an internal source of release. Usually, any internal source of release is limited to a specific enclosure within the pressurized apparatus. This part is known as the containment system.

Static Pressurization

The technique of static pressurization allows for the design of apparatus to be purged and pressurized outside the hazardous area and then taken into the hazardous area without a supply of the protective gas connected. It seems likely that this technique will have very limited application.

Various safeguards are required to ensure that the pressurization has been successfully completed, and that the required overpressure is maintained.

The technique cannot be used where there is an internal source of release.

The main requirements are shown in Figure 10.1.

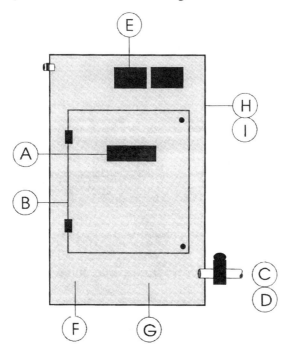

Figure 10.1   *General Requirements for Static Pressurization*

KEY FOR FIGURE 10.1

A   Label warning not to be opened in a hazardous area.

B   Covers must require tool to open.

C   The enclosure must be filled with inert gas whilst in the non-hazardous area.

D   Filling method shall ensure that the initial atmosphere of the enclosure is diluted at least 100 times by the protective gas.

E   Two automatic devices to operate if the overpressure falls below the minimum value specified by the manufacturer. (Devices may be used to isolate electrical power to apparatus, or alarm etc. at the discretion of the user; see earlier text.)

F   No internal sources of release of hazardous substances permitted.

G   Minimum overpressure at pressurization defined by greater of:

   i   Greater than the maximum pressure loss in normal service over 100 times the time taken to ensure capacitors discharged to safe level and temperature of internal surfaces meet T-class specified. (Minimum time of 1 hour)

   ii   50 Pa above atmospheric pressure at place of installation.

H   Enclosure to withstand 1.5 times maximum operating pressure with minimum of 200 Pa.

I   Enclosure to provide at least IP40 protection.

In practice, the technique of static pressurization seems difficult to apply to all but a few special circumstances. Since the static pressurization is a new concept, appearing for the first time when Edition 2 of the CENELEC Standard is published, there is not yet any experience on likely certification uses.

Pressurization with Leakage Compensation

Where the enclosure cannot be totally sealed, a supply of protective gas is continuously applied in sufficient quantity to compensate for any leakage from the enclosure or its associated ducts.

The main requirements are explained by reference to Figure 10.2.

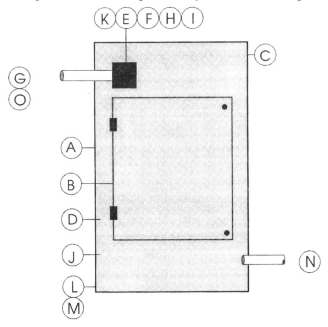

Figure 10.2 *General Requirements for Pressurized Enclosure with Leakage Compensation*

KEY TO FIGURE 10.2

A   The enclosure must not be adversely affected by the gas used for purging.

B   Doors and covers to either have fasteners requiring a special tool, or be electrically interlocked to any apparatus inside the enclosure which is itself not Ex __ protected.

C   Hottest part of external surface defines T-Class.

D   Precautions to be taken to prevent external hazardous atmosphere coming into contact with internal surfaces which exceed stated T-Class if pressurization fails, until surface has cooled to a temperature less than the T-Class. (May require time dependent interlock)

E   Safety device for monitoring the flow of protective gas; timers, relays etc. must be provided.

F   The device(s) stated in E must, if located within the purged enclosure or elsewhere in the hazardous area, not be capable of causing ignition. This will mean that they must be protected by another method of protection. Alternatively, they may be located outside the hazardous area.

G   Volume of protective gas used for initial purging to be such that, after purging, the concentration of flammable gas within the enclosure shall be less than that which equates to 25% by volume of the LEL concentration. [2] Although a figure of five times the volume of free space is often quoted, this may not be sufficient to achieve the 25% by volume of LEL requirement.

H   Flow of protective gas to be monitored at outlet.

I   Device(s) to be provided to either alarm or disconnect (at the discretion of the user and dependent on the zone of use) if the overpressure falls below the minimum level set by the manufacturer.

J   Minimum overpressure of 50 Pa relative to the ambient pressure outside the enclosure.

K   Safety valve to limit internal overpressure.

L   Enclosure to provide at least IP40 protection.

M   Pressurized enclosure to withstand 1.5 times the maximum overpressure stated by the manufacturer, with a minimum of 200 Pa. (The value is normally 1.5 times value of safety valve.)

N   Input for protective gas.

O   Outlet to have device to protect against incendive sparks reaching outside, if required by Table 10.1

Apparatus Group	Zone at exit of purge exhaust	Pressurized apparatus including parts capable of producing incendive arcs and sparks in normal service	Pressurized apparatus containing no normally arcing or sparking parts
I	Non-hazardous Hazardous	Not required Required	Not required Required
II	Non-hazardous Zone 2 Zone 1	Not required Required Required	Not required Not required Required

Table 10.1   *Requirements for Spark Arrestors at Purge Outlet According to Outlet Zone and Contents of Enclosure*

Pressurization with Continuous Flow

This technique is the most complex of the three types of pressurization recognised by Edition 2 of the CENELEC Standard EN 50 016, and requires the pressurized enclosure to maintain an internal overpressure with a continuous flow of protective gas through the enclosure. The technique is usually used for cooling of hot parts within the enclosure.

The general principles are similar to those described in Figure 10.2, except that the device(s) 'I' will be set to operate on the minimum flow rate.

Apparatus with an Internal Source of Release of Hazard (Dilution)

Where the pressurized apparatus contains an internal source of release, the 'area' (normally three dimensional) around the source where the hazard will exist at concentrations which are above a 'safe level', [2] is termed the dilution area. This area is normally within a specific part of the enclosure, for example a flame analyzer etc. The mechanism by which the hazard enters the pressurized enclosure (sample pipe etc.) is known as the containment system.

No ignition capable apparatus (sparks or hot surfaces) is allowed within the dilution area.

The Standard considers three possibilities for the containment system:

a) where the containment system is wholly enclosed such that there is no intentional leakage into the pressurized enclosure.

b) where there is limited release from the containment system.

c) where there is unlimited release from the containment system. [3]

The general design of the pressurized apparatus should be such that any release of hazardous substance (from the containment system) should be located near to the protective gas outlet, and any ignition capable electrical parts located near to the protective gas inlet.

If the containment system is totally enclosed, for example a metallic pipe with no moving joints, and glass to metal seals, then the containment system may be deemed to be infallible, which is to say there will be no source of release and such release need not be considered further.

If the containment system has a limited or unlimited source of release, and if the technique of pressurization with leakage compensation is used, the protective gas must be an inert gas; air is not permitted. Inert gas is also required if unlimited sources of release exist and the technique of continuous flow (continuous dilution) is employed.

The details of pressurization where there are internal sources of release are complex, and those requiring more detailed knowledge should consult the current edition of the Standard. It is also likely that there will be considerable differences in approach between certification bodies, and no doubt there will be extensive guidance in the form of interim guides and amendments to the second edition if it becomes published in its existing form.

Linked Enclosures

There are situations where it is desirable to link a number of items together via one source of protective gas.

The second edition of the European Standard makes provision for situations where the supply of gas is common to a number of enclosures, but does not define precisely how this is to be achieved.

The linked arrangement shown in Figure 10.3 can be used to explain the main requirements, although, because of the arrangement of enclosures, would not be applicable for some conditions.

The safety device 'A' may be common for all enclosures, but must be such as to consider the worst possible condition of any or all enclosures.

The opening of one enclosure in the chain does not necessitate removal of power to all the enclosures if:

a)   The supply to the enclosure to be opened is isolated

and b)   The safety device will still monitor the flow through the other enclosures.

and c)   The correct start-up sequence is used for the enclosure which has been opened, before it is powered up.

150

It is difficult to see how these conditions can be met in a series arrangement as shown, unless some additional sensing to each enclosure is provided. On the other hand, again without some additional sensing devices, it is difficult to see how a parallel arrangement will allow the worst case (for example a blockage in the input of one enclosure) to be successfully monitored, and before any enclosure in a parallel arrangement could be opened, there would presumably need to be an isolation of the supply of protective gas to that enclosure to prevent pressure loss to the other enclosures.

In practice, if one enclosure is opened, the power to all enclosures will be removed.

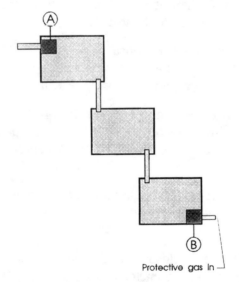

Protective gas in

Figure 10.3   *General Principles of Interlinked Enclosures*

KEY TO FIGURE 10.3

A   Sensor for minimum over-pressure

B   Sensor for high over-pressure.

Use of Component Certified Control Systems

In recent years, it has become quite common for an intermediate manufacturer or an end user to purchase a Component Certified control system which he can then attach to an uncertified enclosure. As always with Component Certificates, the conditions of use need to be carefully established, and the complete apparatus should be certified to give a Certificate of Conformity (CENELEC). The use of the Component Certified control system means that this aspect will not be further considered in the overall assessment, providing that the control apparatus is operating within the scope of its Component Certification.

Layout of Apparatus and Components Within Enclosures

Apparatus within the pressurized enclosure needs to be arranged in such a way as to avoid possible gas traps. In addition, if there are energy storing components such as capacitors, or hot spots within the enclosure, then there may be a restriction on the opening time of the enclosure after the power has been disconnected, or a requirement for the addition of discharge resistors with capacitors, or a combination of both.

General Constructional Requirements

Because of the sealing and structural requirements for pressurized enclosures, a normal IP rated enclosure may not be sufficient, and thus enclosures are normally purpose made. Within the requirements it is, however, quite possible to produce pressurized apparatus which includes push buttons, switches and even full QWERTY keyboards, providing the impact test requirements of EN 50 014 can be met.

Marking

The normal requirements of EN 50 014 apply, namely:

The name of the manufacturer or his registered mark.

The manufacturer's type identification.

The symbol EEx p

The symbol for the group of the electrical apparatus, for example IIA, IIB, IIC.

The temperature class (T1 to T6).

A serial number.

The identification of the testing station.

The certificate number, commencing with the year of certification.

The letter 'X' at the end of the certificate number if there are special conditions of safe use; for example if the pressure and/or flow limiting devices are not included as part of the apparatus.

The point or points at which pressure is to be monitored. (This information can be included in the associated documentation instead)

In addition, type of protection 'p' requires:

* The internal free volume of the apparatus.

* The type of protective gas if other than air.

* Purging details of:
    minimum purge flow rate
    minimum purge duration

* The minimum and maximum overpressure.

* The maximum leakage rate from the enclosure.

Plus, in addition if appropriate

* Minimum additional purge duration per unit volume of ducting.

* Temperature or range of temperature for the protective gas inlet if required by the manufacturer.

  That the enclosure is protected by static pressurization.

  A warning that the enclosure shall be filled only in the non-hazardous area. (For static pressurization)

  A warning 'Do not open in the hazardous area' (Static pressurization)

  A warning label if an inert gas is used which could present a danger of asphyxiation.

Plus, where there is an internal source of release:

* The category of internal release.

* The minimum flow rate of protective gas (unless the source of release in infallible).

* The maximum inlet pressure to the containment system.

* The maximum flow rate into the containment system.

* The maximum oxygen concentration in the containment system.

* A warning concerning the enclosure containing an inert gas and that the enclosure may release a flammable substance. (applies if inert gas is used as the protective gas)

* Not applicable to static pressurization.

Notes and References

1 Pressurization with continuous flow is primarily used for cooling hot spots; eg thyristors and transformers etc. Continuous flow should not be confused with dilution, which is primarily for reducing hazardous releases inside the enclosure to a safe level.

2 For the purposes of pressurization, if the protective gas is air, a safe level is considered to be less than 25% of the LEL of the flammable source of release. If the protective gas is an inert gas, it is possible to achieve a safe level by operating at a level which ensures that the oxygen content is less than 50% by volume of the concentration of oxygen at which the UEL occurs.

3 For the purposes of the Standard, the only sort of unlimited release which is considered is that release caused by liquids from which a flammable gas or vapour can evolve. Any other unlimited sources of release (that is unlimited releases of gas or vapour) are excluded from the Standard.

CHAPTER 11

Encapsulation & Special Protection

Encapsulation & Special Protection

SUMMARY		
	ENCAPSULATION	SPECIAL PROTECTION
CODE	EEx m	Ex s
STANDARD	EN 50 028 BS 5501: Part 8	SFA 3009
CODE OF PRACTICE	NOT YET PUBLISHED	BS 5345: Part 8
ZONES OF USE	1 & 2	1 & 2 (0)
LIVE MAINTENANCE	No	No
TYPICAL USES	Probably low cost apparatus because repair is difficult	Not established

The dramatic improvement in readily available and easy to use epoxy resins and other casting compounds has led to the increase in popularity of encapsulation as a means of protecting small electrical apparatus for use in hazardous areas. Until 1987, when the CENELEC Standard EN 50 028 was published, encapsulation as a method of protection in its own right was not recognised, and although the technique was commonly used, it could only be awarded the code 's' for special protection to a National Standard. With the introduction of the CENELEC Standard, 'encapsulation', 'm' is now accepted throughout Europe as a method of protection.

Although special protection 's' does not always involve encapsulation techniques, it is doubtful if it warrants a chapter on its own, and thus the main aspects have been included at the end of this chapter.

Main Requirements (Encapsulation)

The main requirement for encapsulation 'm' is that the apparatus to be protected be encapsulated in compound to a depth of at least 3 mm to the free surface. (This may be relaxed to 1 mm for small components that do not have any free surface exceeding 2 cm^2.) A summary of other requirements is explained by reference to Figure 11.1

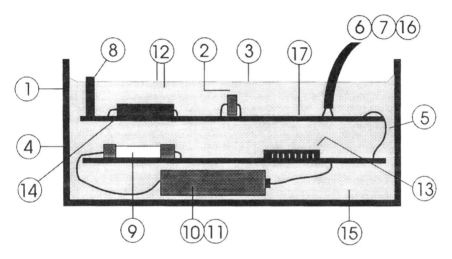

Figure 11.1   *Principles of Encapsulation Protection 'm'*

KEY TO FIGURE 11.1

1   Wall thickness exceeds 1 mm, therefore no minimum compound depth in this direction, unless required for electrical clearances. (see Table 11.1)

2   Minimum depth of compound at least 3 mm. (Minimum depth of at least 1 mm allowed for small components, but since apparatus will not meet 7 Nm impact test an 'X' would be required in the certificate number if this relaxation was used, unless certification includes an outer enclosure.)

3   Apparatus must withstand 7 Nm impact test specified in EN 50 014, Clause 22.4.3.1. unless the apparatus is either

a) intended to be located within an enclosure which will meet those tests. (In this instance the certificate number may bear the suffix 'X' if the enclosure is not fully stated in the application.)

or

b) regarded as a component, in which instance only a Component Certificate will be issued.

4   Case of insulating material.

5   Minimum distance between conductor or component and outside of apparatus must be at least 1 mm. This may be achieved by the case alone, or by a combination of case and compound. (If case is metal, then a 1 mm layer of compound must exist between the case and any conductor/component.)

6   Cable subjected to a pull test of the lower value of:

(20 times cable diameter in mm) Nm

or   5 times mass of encapsulated assembly

for one hour (type test) without any evidence of displacement between cable and component at end of test. (Note: The test shall in no instance be less than 1 N.)

7   Permanently connected cable permitted, even for Group I apparatus.

8   Parts (e.g. printed circuit boards) not fully encapsulated, to be subjected to evaluation of adhesion of the compound to ensure that no visible separation is detectable after thermal type tests.

9   Fuse elements must be of enclosed type: for example glass or ceramic cartridge.

10 Primary or secondary cells must be of a type which will not release gas or electrolyte or produce excessive temperature rise in normal operation and under conditions specified by the manufacturer.

11 Expansion characteristic of cells may need to be considered to prevent excessive pressure being applied to the compound. Such problems may be overcome by the use of a layer of suitable flexible elastomer around the cell. (Note: It is important to use an elastomer which may be compressed within its own initial volume since it will be totally surrounded by solid compound.)

12 Surface temperature not to exceed appropriate T-Class, and temperature of compound not to exceed the continuous rated operating temperature.

13 The integrity of protection method 'm' must be maintained even in the case of recognised overloads and any single internal electrical fault (plus sequential faults occurring as a result of any single fault). (But see 14)

14 Film type resistors, single layer helical wire wound resistors, single layer helical coils, plastic foil capacitors, paper capacitors, ceramic capacitors, used at not more than 2/3 of their rated voltage and power are exempt from consideration of fault conditions.

15 The encapsulation must not have voids, except those present in components, for example opto-isolators and relays, where voids up to a maximum of 100 cm^3 are permitted if the thickness of compound surrounding the components is at least 3 mm. Where voids are less than 1 cm^3, the thickness surrounding the component may be reduced to 1 mm.

16 The circuit must be capable of withstanding a prospective short circuit current of 4000A. (This probably means that a suitable

ceramic sand filled fuse will be required at the input.) If this condition is not met, the marking for the apparatus shall include the letter 'X'.

17 Faults between conductors on a printed circuit board need not be considered if the distance between conductors is at least that given in Table 11.1.

Distances Between Electrical Conductors

The possible fault conditions resulting from shorting two conductors of

the same circuit,

the circuit and any earthed metallic part,

two separate circuits,

need not be considered if the conductors are mechanically fixed with respect to one another prior to encapsulation, and the distance between them is at least that shown in Table 11.1. (The values quoted for Rated Voltage are regarded as nominal values, and may be exceeded by 10%.)

Rated Voltage for the insulation V rms	Minimum Distance mm
380	1.0
500	1.5
660	2.0
1000	2.5
1500	4.0
3000	7.0
6000	12.0
10000	20.0

Table 11.1   *Minimum Distances through Compound to Exonerate Consideration of Faults between Live Parts*

Specific Requirements for Type of Compound

The Standard does not make any direct requirements for the type of compound to be used, and allows any thermosetting, thermoplastic, elastomeric material, including or excluding fillers or additives. Once any of these have solidified, they are considered to be a compound suitable for encapsulation 'm'. However the requirements for the certification type tests are quite onerous, and may result in some otherwise suitable compounds from being excluded.

Type Tests for Compounds

Electric Strength

A solidified disc of compound (50 mm +/- 2 mm diameter, 3 mm +/- 0.2 mm thick) must withstand a 4kV, 48-62 Hz applied voltage via 30 mm diameter electrodes placed either side at the centre of the disc for five minutes. The test is required to be carried out at the highest operating or storage temperature allowed for the compound in its use as encapsulation 'm'.

Thermal Tests

In addition to tests to ensure that neither the T-class nor the maximum temperature of the compound are exceeded, a complex (and expensive) thermal cycling test is required. This test involves a series of temperature changes, starting at room temperature, then an elevated test at the maximum permitted ambient temperature +40°C, then after stabilisation, the apparatus is fully energized, and again allowed to stabilise in this condition. After stabilisation, power is removed, and the apparatus returned to normal room temperature. After stabilisation in this condition, a reduced temperature of 5°C less than the minimum permitted compound temperature is applied, with three stabilising power on / power off cycles before being allowed to return to room temperature.

The test is considered satisfactory if no visible signs of deterioration, such as cracks, shrinkage etc. which could impair the integrity of the method of protection 'm', are observed.

Marking: EEx m

Apparatus shall be marked with the following:

The name of the manufacturer or his registered trade mark

The manufacturer's type identification

The symbol EEx m

The symbol for the Gas Group (IIA, IIB, IIC)

The temperature class (T1-T6)

The indication of the testing station

The certificate number

Input and output electrical data (voltage, current etc.)

Plus, if appropriate

The suffix X if use has been made of the relaxation on compound depth to 1 mm for small components.

The suffix X if the apparatus is for connection to an external supply, and cannot withstand a prospective short circuit current of 4000A, and the permitted maximum prospective short circuit current of the supply.

Information on any external fuse required by the certificate.

Special Protection

Special protection 's' is not the subject of a CENELEC Standard. It exists to allow the certification of apparatus and equipment which does not comply with any of the recognised methods of protection, but which, in the opinion of the Certification Authority, (usually after extensive testing) is safe for hazardous area use.

In the UK, special protection is covered by the BASEEFA Standard SFA 3009. The code is Ex s, and apparatus certified to the Standard is generally acceptable for zone 1 and zone 2 use. Theoretically, special protection may be certified for zone 0, but this is extremely rare, and usually means that apparatus is in fact intrinsically safe to category 'ia', but for some technical reason will not conform to the Standards for intrinsic safety.

Special protection certification has also been used to allow apparatus with a method of protection suitable for zone 2 only to be used in zone 1 in the UK. An example of this might be some oil immersed apparatus containing only normally non-sparking circuits. Such apparatus would, in any case, be acceptable for zone 1 use outside the UK by virtue of its EEx o certificate, but until the CENELEC Standard for oil immersion is revised (see Chapter 10) would need an additional assurance, given by an Ex s Certificate of Assurance for zone 1 use in the UK.

Marking: Ex s

Special protection apparatus shall include the following marking.

Manufacturer or agent

Type identification

The Standard number; e.g. SFA 3009

The BASEEFA mark

164

The certificate number

The code Ex s

The gas group

The temperature class

The ambient temperature range if other than -20°C to +40°C

Any other marking required by the certificate

CHAPTER 12

Increased Safety

Increased Safety

SUMMARY	
CODE	Ex e EEx e
STANDARDS	BS 4683: Part 4 BS 5501: Part 6 EN 50 019
UK CODE OF PRACTICE	BS 5345: Part 6
ZONE OF USE	Zones 1 and 2
LIVE MAINTENANCE	No
TYPICAL USES	Rotating machines, lighting, junction and terminal boxes

Introduction

Increased safety, code letter 'e', assures safety against electrical ignition by applying measures to ensure, with a minor degree of security, that the possibility of excessive temperatures and the occurrence of arcs and sparks in the interior and on external parts which do not produce them in normal service are prevented.

So, increased safety, like N-type protection, achieves its safety by removing possible sources of ignition. However, the level of security is greater than for Ex N, since fault conditions are considered.

The technique of increased safety has only recently become popular in the UK, but in Germany it has been one of the most widely used methods of protection for some time. Increased safety is not recognised as such in North America, although in some respects the North American technique of Non-Incendive protection is similar to increased safety.

In the UK, prior to the publication of the CENELEC Standards, BS 4683: Part 4 was used as the certification Standard. There was also a BASEEFA Standard SFA 3008 to which apparatus could be certified.

Main Constructional Requirements (EN 50 019)

Ingress Protection

Enclosures containing live bare parts must provide protection of at least IP54. Enclosures containing only insulated parts shall provide protection of at least IP44.

Relaxation of the above requirements may apply in certain instances to motor starting resistors, and electrical machines installed in clean rooms.

Avoidance of Frictional Sparks

Design measures must be taken to avoid the possibility of frictional sparks, for example between the stator and rotor of a motor. The requirements are stated in Table 12.1, and apply to radial air gaps measured at standstill.

Number of Poles	Air Gap (mm) in relation to diameter of rotor D (mm)		
	$D \leq 75$	$75 < D \leq 750$	$D > 750$
2	0.25	$0.25 + \dfrac{D-75}{300}$	2.7
4	0.2	$0.2 + \dfrac{D-75}{500}$	1.7
6 and more	0.2	$0.2 + \dfrac{D-75}{800}$	1.2

NOTES

1   If the length of the core L exceeds the value of 1.75D, the value calculated for the required air gap should be multiplied by $L/(1.75D)$.

2   For machines supported by one or more sleeve bearings, the calculated value shall be multiplied by 1.5

Table 12.1   *Minimum Air Gap for Machines with Roller-Element Bearings*

Mechanical Strength

Apparatus is required to comply with the mechanical strength tests of EN 50 014. These requirements are summarised in Table 12.2. Testing should normally be carried out at 20°C ± 5°C. The test apparatus is as described in Appendix 9.

Group	Impact Energy E joules			
	I		II	
Risk of mechanical damage	High	Low	High	Low

Guards, protective covers, fan hoods, cable entries
Plastic enclosures
Light metal or cast metal enclosures
Enclosures of other materials:
  wall thickness less than 3mm for Group I
  wall thickness less than 1mm for Group II
	20	7	7	4

Light transmitting parts without guard	10	4	4	2
Light transmitting parts with guard (tested without guard)	4	2	2	1
NOTE. Apparatus tested to the values of low risk of mechanical damage require an 'X' in the marking.				

Table 12.2   *Mechanical Strength Tests for Electrical Apparatus (EN 50 014)*

Ignition from Hot Surfaces

The requirements for increased safety lay great emphasis on the precautions to avoid the possibility of ignition from hot surfaces.

The Standard refers to the term 'limiting temperature', which is defined as the highest permissible temperature of an electrical apparatus or part of an electrical apparatus. The limiting temperature is the lower of

a)   The maximum surface temperature associated with the appropriate T-class.

or b)   The temperature determined by the limit of the thermal stability of the materials used.

The temperatures concerned must include the conditions of starting, normal operation, and any recognised overload at the end of time t_E (t_E is the time taken for ac windings, when carrying the starting current (I_A) to be heated up from the temperature reached in rated service and at maximum ambient temperature to the 'limiting temperature'.)

Rotating Machines[1]

For rotating machines, the length of the time t_E shall be such that when the rotor is locked, the motor can be disconnected by a current dependent protective device, before time t_E has elapsed. The type of protective device shall be indicated by marking on the machine. In no case shall the time t_E be less than 5 seconds. In other words, Ex e motors must be fitted with a protective device to prevent excessive temperatures under stall or locked rotor conditions.

The disconnection device shall be either

a)   installed in the winding

or b)   be an external current dependent time lag protective device. (Fuse)

The latter type of device is generally only suitable for motors which are subject to continuous service involving easy (light or no load) starting and infrequent starting.

It should be noted that external current dependent time lag protection devices are not normally certified, and thus this aspect requires most careful consideration by the user. A tripping device of 20% less than the t_E time is recommended both by EN 50 019 and by the UK Code of practice, BS 5345: Part 6, to allow for a possible margin of error. These devices should be mechanically robust, and suitable 'in all respects' for the environment of installation. This requires the device itself to have some method of protection if it is to be located in a hazardous area. After all, there is little point in disconnecting a motor so that hot surfaces are prevented, if the act of disconnection produces an arc which is incendive!

In order to ensure that insulation is not unduly stressed by temperature, it is normal to de-rate the maximum permissible temperatures of the different classes of insulating material as shown in Table 12.3

	Class of insulation				
	A	E	B	F	H
Maximum permitted temperature (limiting temperature) °C at end of time t_E	160	175	185	210	235

Table 12.3 *Maximum Insulation Temperatures*

Assuming a maximum ambient temperature of 40°C, this corresponds to maximum temperature rises for each class of insulation as follows:

Class A	120°C
E	135°C
B	145°C
F	170°C
H	195°C

This reduction of permitted temperature (10%) for the insulation class means that increased safety motors are larger than the equivalent motor for non-hazardous area use for the same kW rating.

Other Electrical Aspects and Other Apparatus

Luminaries for Mains Supply

Lighting units are often protected using increased safety. The following types of mains operated light sources are acceptable:

Fluorescent lamps of cold starting type with single pin caps.

Filament lamps for general lighting service.

Mixed light (MBT) lamps

Other lamps where there is no danger that parts of the light source can reach temperatures higher than the limiting temperature following breakage of the bulb.

Lamps containing free metallic sodium are not permitted.

Minimum distances between a lamp and its protective cover are required to be in accordance with Table 12.4

		Minimum Distance in mm
Fluorescent tubes	with protective cover	5
	with protective outer tube	2
Other lamps	wattage < 60W	3
	60 - 100 W	5
	100 - 200 W	10
	200 - 500 W	20
	>500 W	30

Table 12.4   *Minimum Distances between a Lamp and its Protective Cover*

Lampholders, together with the lampcap are required to comply with the flame transmission safety levels of flameproof protection 'd' for Gas Group IIC (or Group I for mining applications) unless they can be shown to achieve safety by other means. In addition,

172

precautions must be taken to prevent self loosening of lamps with screw caps.

Measuring Instruments and Measuring Transformers

Measuring instruments and transformers are required to be able to continuously withstand 1.2 times their rated current or voltage without exceeding the limiting temperature.

Measuring instruments with moving coils are not permitted unless the coils have leads which are not likely to break. This is interpreted as being where movements of not more than $1°$ or 0.5 mm are involved.

Current transformers and current carrying parts of measuring instruments shall be capable of withstanding the currents shown in Table 12.5 without any reduction in their safety against explosions.

	Current carrying parts of measuring instruments	Current transformers
I th (b)	$\geq 50 \times I_N$	$\geq 100 \times I_N$
I dyn (c)	$\geq 1.3 \times 125 \times I_N$ (a)	$\geq 1.3 \times 250 \times I_N$

NOTES

a   The factor of 1.3 is a safety factor.
b   The currents are applied for 1 second.
c   The currents are applied for 0.01 seconds minimum.

Where: I th is the Thermal Current Limit - the value (rms) of the current by which the conductor is, within one second, heated up from the temperature reached in rated service at maximum ambient temperature to the limiting temperature.

I dyn is the Dynamic Current Limit - this is the peak value of current the electrical apparatus can sustain without damage.

I_N is the rated current.

Table 12.5   *Current Carrying Requirements for Transformers and Measuring Instruments*

The implication of Table 12.5 is that an instrument intended to operate at 20mA, must be able to withstand 50 x 20mA = 1 amp for 1 second without exceeding the limiting temperature. In no case shall temperatures in excess of 200°C be attained. Additionally, it shall be able to withstand at least 20mA x 1.3 x 125 = 3.25A for at least 0.01 seconds.

Terminals for External Connection

Terminals must be 'generously dimensioned' to permit the effective connection of conductors of cross section at least corresponding to the rated current I_N. The terminals shall be fixed in their mountings without the possibility of self-loosening, be such that they cannot slip out from their intended location and ensure proper contact with the conductors without deterioration of the conductors even if multi-strand conductors are used in terminals intended for direct clamping to the cores. Insulating materials must not be used to transmit contact pressure.

Internal Connections

The following methods are allowed:

Screwed or bolted connections locked against loosening

Crimped connections

Soldered connections where the conductors are also mechanically supported

Brazing

Welding

Clearances, [2] Creepage, [3] and Insulation Tracking[4]

Clearance distances between live conducting parts must comply with the minimum values stated in Table 12.6 to be considered non-sparking.

Rated voltage for the insulation required	Minimum clearance in mm
60	3
250	5
380	6
500	8
660	10
1000	14
3000	36
6000	60
10000	100

Table 12.6   *Minimum Clearance Distances between Live Conducting Parts*

Creepage distances are dependent on the comparative tracking index[4] of the insulating material, and the rated voltage. The requirements are given in Table 12.7.

Rated voltage for insulation V	Minimum creepage distance in mm			
	Grade of insulation			
	a	b	c	d
30	3	3	3	3
60	3	4	5	6
250	6	8	10	12
380	8	10	12	15
500	10	12	15	18
660	12	16	20	25
1000	20	25	30	36
3000	45	60	75	90
6000	85	110	135	160
10000	125	150	180	240
The rated voltage of the apparatus may exceed the values stated in the table by 10%				

Table 12.7   *Creepage Distances According to Insulation*

Appendix 3 gives details for the measurement of creepage distances for the various shapes and styles of surface.

Solid insulation materials should have mechanical characteristics suitable for temperatures of 20K above the temperature reached in continuous rated service, and should be suitable for temperatures of at least 80°C. Where insulating parts are made of plastics or of laminated material, they must be covered with insulating varnish which has at least the same grade of tracking resistance if the moulded skin on the surface is damaged or removed during manufacture, unless damage as indicated will not adversely affect tracking resistance, or undamaged parts meet the prescribed creepage distances.

Windings must be covered with at least two layers of insulation unless enamelled wires comply with IEC 182-2 Grade 1 meeting additional test requirements or comply with IEC 182-2 Grade 2. Where impregnation of a winding is required by the Standard, impregnation is to be by dipping, trickling or vacuum impregnation. Coating by painting or spraying is not recognised as impregnation. The intention of any impregnation is to ensure that the spaces between conductors are filled as completely as possible, and that good cohesion between conductors is achieved.

Unless windings comply with the requirements of an alternative method of protection such as encapsulation ('m') or intrinsic safety ('ia', 'ib'), wires with a nominal diameter of less than 0.25 mm are not allowed.

Marking

In addition to the normal requirements [5] for manufacturer, protection code etc., as listed below:

The name of the manufacturer or his registered mark

The manufacturer's type identification

The symbol EEx e

The symbol for the group of the electrical apparatus:
I for mines susceptible to firedamp
II or IIA or IIB or IIC for other applications

The temperature class (T1 to T6) or the maximum surface temperature in °C, or both.

A serial number, except for small components, bushings etc.

The indication of the testing station

The certificate number, commencing with the year of certification.

The letter X at the end of the certificate number if there are special conditions for safe use.

increased safety apparatus shall also bear the following information:

Rated voltage and rated current

*   Starting current ratio I_A/I_N

*   Time t_E

**   Thermal current limit I_{th}

**   Dynamic Current Limit I_{dyn}

*** Technical data of lamps, including electrical rating and dimensions

Any restrictions in use

Special protection devices required

KEY:

*   For rotating machines and a.c. magnets.

**   For measuring instruments and measuring transformers.

*** For luminaries.

Notes and References

1 Rotating machines may also comply with BS 5000: Part 15. This Standard, however, refers to EN50 019 etc. for certification.

2 Clearance

The shortest distance through the air between two bare conducting parts.

3 Creepage

The shortest distance between two conducting parts along the surface of the insulating parts.

4 Comparative Tracking Index. CTI

See IEC publication 112 for measurement method.

5 For examples of marking, see Appendix 1

CHAPTER 13

Intrinsic Safety

Intrinsic Safety

SUMMARY	
CODE	Ex ib EEx ib Ex ia EEx ia
STANDARDS	BS 1259 SFA 3012 SFA 3004[1] EN50 020 BS5501: Part 7 EN50 039[2] BS5501: Part 9 IEC 79-11 IEC 79-3
CODE OF PRACTICE	BS 5345: Part 4
ZONES OF USE	'ib': 1 and 2 'ia': 0, 1 and 2
LIVE MAINTENANCE	Yes
TYPICAL USES	Instrumentation, control circuits, other low power apparatus.

Intrinsic safety is probably the most popular method of protection, and, where the required limitations of voltage and current allow its use, is often selected in preference to other methods of protection. It is also the only method of protection 'ia' which is permitted in zone 0. Although it gives a much greater margin of safety than other methods of protection, it enjoys some very specific freedoms and relaxations. Perhaps most notable among these are:

a) the freedom to use certain uncertified apparatus in an intrinsically safe circuit

b) the possibility of live maintenance

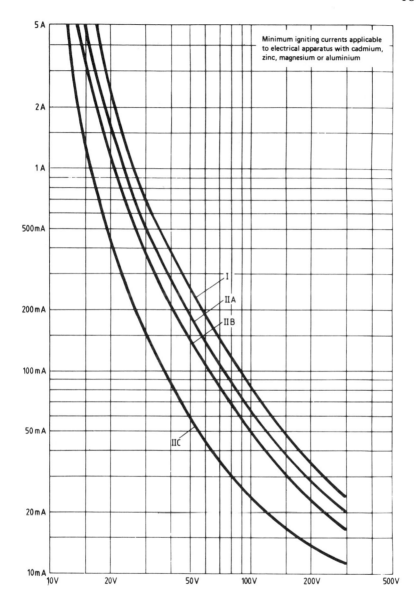

Figure 13.1 *Current/Voltage Curves for Intrinsic Safety (Resistive Circuits)*

c) less restriction on types of interconnecting cable, since mechanical protection of the wiring is less important than for other methods of protection.

Because of its inherent low power, the technique is ideal for instrumentation circuits, local LCD display units, signal conditioning and signal multiplexing.

Basic Principles

Unlike other methods which rely on mechanical means such as enclosures to prevent electrical ignition, intrinsic safety achieves its protection by limiting the electrical parameters to less than the ignition energy level of the hazard. The minimum ignition energy of the test gases used for each gas group have been established experimentally using spark test apparatus, and the data is available in graphical form. (Figure 13.1) It is normally possible to assess the safety of intrinsically safe apparatus by reference to these curves without the need to carry out additional tests.

It will be appreciated that until the power (voltage and current) available to the apparatus has been defined, it is not possible to carry out an evaluation to the curves. Thus, with intrinsic safety there is always a need for some interface unit to define the power available to the intrinsically safe apparatus. (Figure 13.2)

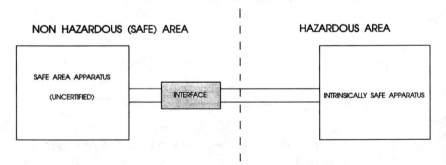

Figure 13.2 *Interface Requirement for Intrinsic Safety*

183

This unit is normally a discrete device such as a galvanic isolator or zener barrier, but it may sometimes be incorporated into an item of safe area apparatus. (Figure 13.3)

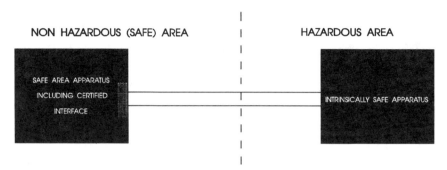

Figure 13.3 *Interface Incorporated into Safe Area Apparatus*

The disadvantage of this latter approach is that (at least some part of) the safe area apparatus must be certified. However, on the other hand it has the possible advantage of saving the cost of installing a separate interface unit. Since the interface unit normally provides a ready means of terminating the field cables and gives a defined point at which intrinsic safety starts, and at which tests and calibration can be carried out, interfaces which are incorporated into the safe area apparatus are not popular: indeed, some end users insist on a separate interface anyway, even where there is already a safety interface built into the safe area apparatus.

Since the interface unit defines the maximum output parameters which will be available from the safe area to the intrinsically safe circuit, even under fault conditions, the interface unit is of major importance to intrinsic safety. Interface units need careful certification, and bear the code

[EEx ia]   (or sometimes [EEx ib])

The inclusion of the brackets signifies that the interface unit is associated safe area apparatus, and should not be installed in the

184

hazardous area unless it is protected by another method of protection, for example, locating it in a flameproof enclosure.

Interface Units: the Zener Safety Barrier[3]

The interface unit may take a number of forms, but is frequently a zener safety barrier or a galvanic isolator. Zener barriers (Figure 13.4) perform their safety function by limiting the current which can pass to the hazardous area via an infallible current limiting resistor at a voltage defined by a zener clamp.

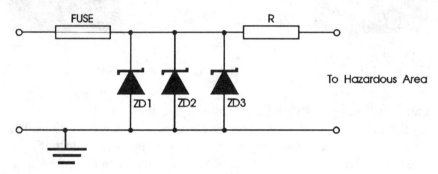

Figure 13.4 *Basic Circuit of a Zener Safety Barrier*

The zener barrier is often referred to in terms of its output safety rating: ZD3, R. For example a 28V, 300 ohm barrier. By examination of the output characteristics and the energy curves, it becomes apparent how intrinsic safety is achieved.

Assume that the zener diodes are rated at 28V maximum, including tolerance. To this value it is necessary to make an adjustment to allow for the effect on zener diodes under certain conditions of elevated temperature which can cause the zener voltage to increase slightly. This adjustment value will normally be established by test, but may be about 10%. Adding 10% (2.8V) to 28V gives 30.8V, and the resistance curves are entered at this value. For gas group IIC, a safe current of 140 mA is indicated. Intrinsic safety requires a factor of safety of 1.5, and thus the current level to operate on is 2/3 of 140 mA = 95 mA (approx). Ohm's law indicates that at

28V, this maximum current will be achieved with a resistor of just under 300 ohms. This resistance value is, of course, the minimum value to assure intrinsic safety conditions are met, and thus a 300 ohm resistor, +/- 1% will suffice.[4]

It should be noted that for lower voltages, the value of the current limiting resistor is usually determined not by the allowable instantaneous short circuit current, but by a value which permits an inductance level which is not restrictive to cable choice.

The end user should note that the resistance presented to a circuit which includes zener barriers is normally about 40 ohms more than the value of the safety resistor. This is due to the resistance of the fuse, and some additional resistance (not shown in Figure 13.4) included in the circuit for the manufacturer's test purposes.

The fuse in the circuit is used to define the maximum current and hence the wattage rating for the zener diodes, since for interface units and other apparatus which, it is assumed may be inadvertently connected to a source of mains supply, the assumption is that a prospective current of 4000A may be available. The use of a fuse (normally a ceramic, sand filled cartridge type) allows the current to the zener diode to be defined at a more reasonable level. The fuse rating is normally of little interest to the end user, since most barriers are designed to be 'short circuit proof'. That is to say the maximum output current is limited by resistive effects at the maximum safe working voltage (usually about 1-1.5 volts less than the rated zener voltage) to a level less than the fusing current.

Interface Units: Galvanic Isolator

The use of galvanic isolation interface units in place of zener barriers is increasing in popularity. The main advantages of galvanic isolators are:

a) They do not require an earth connection for intrinsic safety purposes.

b) They overcome problems of resistive volt drop at the interface - especially important in some 4/20 mA circuits.

c) They perform a complete process function: for example providing an isolated 4/20 mA signal at the safe area terminals corresponding to a 4/20 mA loop in the hazardous area.

d) They often provide a mains powered version direct. This can be especially convenient where there are just a few loops, and no readily available 24V supply.

On the negative side, they are considerably more expensive than zener safety barriers. They are active devices (rather than purely resistive) and thus have an inherent accuracy error which may need to be taken into account on some sensitive loops. Additionally, they absorb power and thus generate some heat. They are larger, and take up more space than the barrier (but, considering point 'c' above probably not excessively so), and unlike the barrier, which may be used for a number of different functions, the galvanic isolator performs a specific task and the correct unit for that task - for example providing a switching output contact, or providing a repeated 4/20 mA signal - must be used.

The various designs of galvanic isolator differ widely, but essentially there is some transformer isolated supply, often using saturation techniques to limit the output power, and some sensing element (resistor etc.) across which the signal in the hazardous area can be monitored, and some function performed. The signal return is often isolated by an intrinsically safe relay (digital signals) or an opto-isolator / isolation amplifier (analogue signals). The basic principles are shown in Figure 13.5.

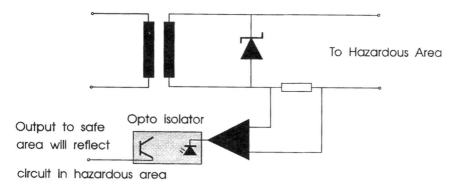

Output to safe area will reflect circuit in hazardous area

Opto isolator

Figure 13.5 *Basic Circuit for Galvanic Isolator (Analogue Signal)*

Categories of Intrinsic Safety 'ia' and 'ib'

There are in fact two methods of protection included under the intrinsic safety umbrella: code 'ia' and 'ib'. The difference between the two concerns the safety achieved under possible fault conditions.

Intrinsic Safety 'ib'

Consider the circuit shown in Figure 13.6, which depicts a basic limiting circuit. Like the zener barrier, the maximum output may be defined by

Voltage of V_{ZD}
Current (instantaneous short circuit) V_{ZD}/R

R

ZD

To Hazardous Area

Figure 13.6 *Basic Voltage/Current Limiting Circuit*

188

If the resistor R is of infallible construction, [5] then the first fault which can reasonably be considered is an open circuit fault of the diode. If this were to happen (an unlikely situation in fact, since diodes normally fail to short circuit) then the circuit would not provide any voltage clamp.

Intrinsic safety category 'ib' is required to meet the requirements of intrinsic safety (that is, comply with the energy curves and provide a safety factor of 1.5) **and still be safe with one fault**. By modifying the circuit as shown in Figure 13.7, and adding an additional zener diode, this condition is met.

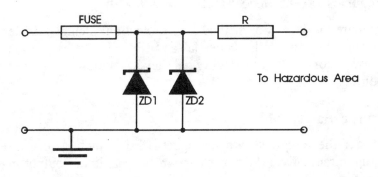

Figure 13.7   *Basic 'ib' Zener Barrier*

Intrinsic Safety 'ia'

Whereas intrinsic safety 'ib' requires safety with one fault, **intrinsic safety category 'ia' requires continued intrinsic safety even with two faults.** [6] In this analysis (referring to Figure 13.7) both zener diodes can now be considered to have failed to open circuit, and thus again there is no voltage limit on the circuit. The solution is to include a third zener diode, giving the resulting circuit for the zener barrier shown in Figure 13.4. [7]

Apparatus and circuits which meet the requirements of intrinsic safety 'ia' are acceptable for use in zones 0, 1, and 2. [8] Intrinsic safety 'ia' is, in practice, the only method of protection which is acceptable for zone 0 use.

Field Apparatus in Intrinsically Safe Circuits

It will be appreciated from the earlier discussion of interface units, that the spark which might be obtained under instantaneous short circuit conditions from the output of the zener safety barrier will not cause ignition. (Figure 13.8)

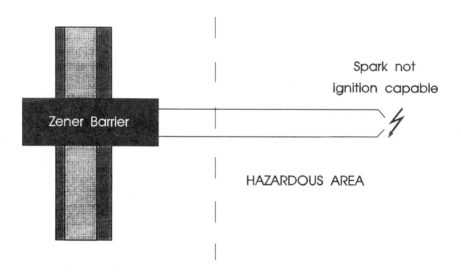

Figure 13.8   *Consideration of Interface Output*

Clearly, the situation does not change if, instead of a spark between the ends of the wires, a switch is introduced into the circuit. In fact, any device which is resistive in nature and thus non energy storing can be added to the circuit and the overall system consideration will still be intrinsically safe. (Figure 13.9)

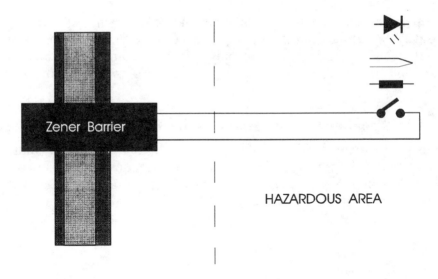

Figure 13.9   *Non Energy Storing Field Apparatus*

Such items are generally termed **simple apparatus** and have a very significant contribution to intrinsically safe circuits, since simple apparatus does not need to be certified. The definition for simple apparatus is perhaps best stated in the UK Code of practice, although similar clauses appear in the CENELEC Standard (EN50 014 clause 1.3) and in IEC 79-11.

Simple electrical apparatus and components may be used in intrinsically safe systems without certification, provided that they do not generate or store more than 1.2V, 0.1A. 20µJ and 25mW in the intrinsically safe system in the normal or fault conditions of the system.

The Code also explains that apparatus which can dissipate not more than 1.3 watts [9] may be awarded a T-Class of T4 without further evaluation. This is convenient, since most interface units, certainly the zener safety barrier, cannot give an output power of this level. (See Figure 13.10)

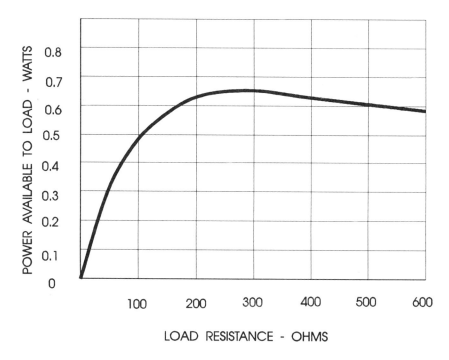

Figure 13.10 *Matched Power Output and Output Curve of Typical Zener Safety Barrier (28V, 300Ω)*

Clearly, if the apparatus in the hazardous area intrinsically safe circuit has some inductance or capacitance, then even with the maximum voltage and current limited to intrinsically safe levels (by the interface unit) there is the possibility that a capacitive discharge or inductive surge could cause ignition. Thus apparatus or components with any appreciable amount of capacitance and inductance is not simple apparatus, and needs to be certified. If the effective capacitance or inductance level is within that allowed for intrinsic safety, [10] the apparatus will be acceptable for intrinsic safety, but will of course need to be a certified Ex ia/ib or EEx ia/ib unit.

Intrinsically Safe Systems

From the foregoing, it will be appreciated that the overall system configuration, normally comprising an interface unit, cables, junction boxes and field apparatus, needs careful consideration where the method of protection is intrinsic safety. Clearly, if some field apparatus contains capacitance and has been certified as being intrinsically safe at a voltage of 28 volts, it may not (in fact almost certainly will not) be safe if connected to a source of potential of say 100V. By considering the simple block diagrams of Figures 13.11 and 13.12, it will be appreciated that in the case of simple apparatus, the simple apparatus itself does not need to be certified, but needs to be connected to a certified [EEx ia/ib] or [Ex ia/ib] interface to ensure intrinsically safe conditions. In the case of non-simple apparatus however, the apparatus must be certified intrinsically safe. The interface unit is still required, since the maximum power available in the entire circuit must still be controlled.

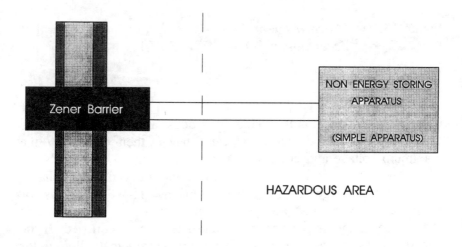

Figure 13.11 *System Comprising Interface and Simple Apparatus*

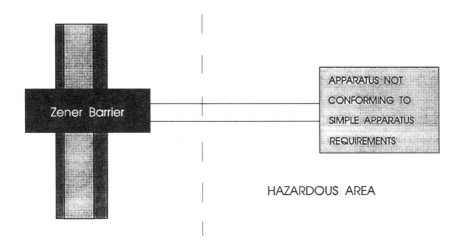

Figure 13.12   *System Comprising Interface and Non-Simple Apparatus*

Sometimes, in addition to the Certificate of Assurance or Certificate of Conformity for the individual items of apparatus, there will also be a System Certificate which defines the essential safety criteria for the whole circuit. Where the field apparatus is simple apparatus, the System Certificate will normally be obtained from the interface manufacturer. Where the field apparatus is not simple apparatus, the System Certificate, stating, amongst other aspects, which interface units should be used, will be obtained from the manufacturer of the field apparatus.

If a System Certificate is not available, check the following aspects, to ensure compatibility for system intrinsic safety.

a) The gas group of the interface unit should be the same or better than the gas group for the installation.

b) The gas group for the field apparatus should be the same or better than the gas group for the installation.

c) Check that the interface unit and the field apparatus are suitably certified for the zone of use.

(If there is a mixture of 'ia' and 'ib' apparatus in the system, then the overall system takes the lowest category; that is 'ib'.)

d) Check that the certified parameters for the field apparatus are greater than the certified output parameters of the interface.

(for example:

Barrier		Transmitter	
U_Z	: 28 V	U_{max}	: 30 V
I_{OUT}	: 186 mA	$I_{MAX\ IN}$	: 250 mA
P_{OUT}	: 650 mW	$P_{MAX\ IN}$	: 1.3 W
L_{EXT}	: 3.1 mH	L_{EQ}	: 10 μH
C_{EXT}	: 0.39 μF	C_{EQ}	: 0.01 μF

would be OK.)

e) If there is a total cable length in excess of about 1 km between interface unit and field apparatus, see notes on cable parameters below.

f) Check that the temperature class of the field apparatus is suitable for the application.

If in doubt, it is always worth asking for a system drawing from the field apparatus manufacturer. This drawing will give useful information about the installation conditions, and will usually list any possible options for interface units etc. An example of a system drawing is given in Figure 13.13.

195

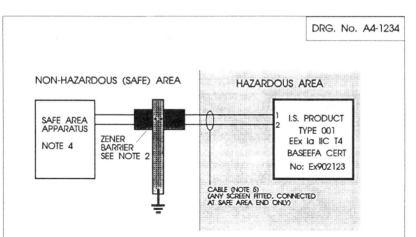

NOTES

1   The installation must conform to the National requirements of the country of use.
For example, UK installations must conform to BS 5345: Part 4: 1977.

2   The zener barrier type MTL 788. Certified (EEx Ia)IIC, Certificate number Ex832452.

3   The electrical circuit in the hazardous area must withstand an AC test voltage of
500V rms test to earth or frame of apparatus for one minute without breakdown.

4   Non-Hazardous (Safe) Area Apparatus. Unspecified, except that it must not
be supplied from nor contain under normal or abnormal conditions, a source
of potential with respect to earth in excess of 250V ac rms or 250V dc.

5   Cable parameters. The capacitance, and either the inductance or the
inductance to resistance ratio of the interconnecting cables shall not exceed
the following parameters:

GAS GROUP	CAPACITANCE μF	INDUCTANCE mH	INDUCTANCE / RESISTANCE RATIO μH/OHM
IIC	0.10	0.37	24
IIB	0.36	1.11	72
IIA	1.01	2.96	192

THE I.S. PRODUCT COMPANY   DRAWING NUMBER A4-1234

SYSTEM DRAWING FOR INTRINSICALLY SAFE I.S. PRODUCT SYSTEM. SYSTEM CERTIFICATE Ex 902124

ISSUE	1	2					DRAWN BY HEXAGON TECHNOLOGY LIMITED
DATE	9.11.90	1.8.92					

Figure 13.13   *Typical System Drawing*

Cable Parameters

Since, with intrinsic safety there are no requirements to use mechanically protected cable, steel wire armoured etc., it is assumed that the cable may be broken, or the field apparatus disconnected whilst the circuits are live. Thus, in addition to the intrinsically safe voltage and current considerations, the capacitance and inductance available at any break in the cable must be considered. These properties are made up of a total of any capacitance or inductance in the field apparatus, plus that of the interconnecting cable. Where the field apparatus is simple apparatus, it has no appreciable capacitance or inductance, and thus the maximum permissible may be taken in the interconnecting cable. (The applicable values will be quoted by the interface manufacturer, either in his installation drawing or on the System Certificate).

Where the field apparatus is not simple apparatus, if there is a System Certificate, or system drawing, then the maximum allowable cable parameters will be specified. If no such documents are available however, use the information on cable parameters supplied by the interface manufacturer (which are applicable to simple apparatus) and subtract the values of 'C_{EQ}' and 'L_{EQ}' which should be stated on the apparatus certificate (Certificate of Assurance or Certificate of Conformity) for the field apparatus to obtain the relevant cable parameters.

It should be noted that cable parameters are often quoted for gas group IIC. The maximum allowable parameters for group IIC may be multiplied by 3 for gas group IIB installations, and by 8 for gas group IIA applications. In practice, unless the interconnecting cables are unusually long (1km or so) there will rarely be any difficulty in complying with the requirements.

Multicore Cables

Multicore cables may be used for intrinsically safe circuits. The main rule, which should never be disregarded, is that intrinsically safe and non-intrinsically safe circuits should never be run in the same multicore.

Within a multicore, each intrinsically safe circuit should occupy adjacent pairs of wires.

If individual circuits within a multicore are screened, the screens should be insulated from one another.

Entity Approval (USA)

In the USA, instead of issuing a system certificate, it is normal to quote a set of Entity Parameters in connection with the interface output and the field apparatus input. By the same simple rules as indicated earlier for system compatibility, and the above explanation of cable parameters, the necessary system safety criteria can easily be established.

Maintenance

One other important aspect of intrinsic safety is that, unlike all other methods of protection, live maintenance, without a gas free certificate or hot work permit, is allowed. This aspect is considered further in Chapter 17.

Marking

Instrinsically safe apparatus certified to the CENELEC Standards, should be marked with the following:

The name of the manufacturer or his registered trade mark

The manufacturer's type identification

The symbol EEx ia or EEx ib (or [EEx ia] or [EEx ib] for associated safe area apparatus)

The symbol for the gas group (IIA, IIB, or IIC)

The temperature class (T1-T6) (Not applicable for associated safe area apparatus)

The identification of the testing station

The certificate number

Plus, if appropriate

Equivalent inductance of circuit: L_{EQ}

Equivalent capacitance of circuit: C_{EQ}

Maximum permissible applied voltage: U_{MAX}

Alternatively, for zener barriers,

The maximum voltage which may be applied to the non-intrinsically safe terminals: U_M

The maximum output voltage of the barrier: U_Z

Notes and References

1   BASEEFA Standard SFA 3004 is only applicable to shunt diode safety barriers. Although obsolete, the document provides useful information for anyone interested in a more detailed explanation of design aspects for zener safety barriers.

2   EN50 039 refers to intrinsically safe systems, as opposed to apparatus.

3   The explanation given here for the design and operation of zener safety barriers is deliberately brief, since the intention is to give a general view of intrinsic safety rather than a detailed analysis. The reader requiring a more thorough treatment should refer to 'Intrinsically Safe Instrumentation: a guide' by the same author, ISBN 0 9508188 0 1, available from Hexagon Technology Limited.

4   At the time of going to press the 10% rule is about to be dropped by the CENELEC Test Houses. This will allow significant improvement (reduction) to the resistance values needed for infallible current limitation.

5   In intrinsic safety design much use is made of components with an established failure mode. If such devices can be used in a way so that if a failure occurs this will produce a result which is intrinsically safe, infallible or inviolate status may be claimed.

For example a suitably rated wire wound or metal film resistor is considered to fail only to an open circuit condition. Thus, used as a series current limiting device, if it fails it will fail to a state where no current passes - the ultimate limit! Thus a wire wound or metal film resistor, of suitable wattage rating is deemed as an infallible component for current limiting, and its failure or fault conditions need not be considered further.

6 The fault count for intrinsic safety 'ia' and 'ib' does not take into account the failure of field wiring.

7 The European Standard EN50 020 (and other National Standards, for example SFA 3004) allow the use of two zener diodes for 'ia' zener barriers under closely specified conditions. These include an elevated temperature test and a pulse test at currents several orders of magnitude higher than the fuse rating which protects the zener diodes in circuit, as well as stringent mechanical/connection requirements. The tests have to be performed on every diode used if the two diode 'ia' route is selected by the manufacturer of the zener safety barrier.

8 In Germany, there is a requirement for intrinsically safe apparatus for use in zone 0 to be galvanically isolated.

9 It has been shown that small components with a surface area size >20 mm^2 <10 cm^2 which dissipate less than 1.3W, will not cause ignition of gases at less than 135°C, and thus the T4 rating may be applied to such items even if not simple apparatus, without the need for testing by the certification authority. Although the local temperature reached on the surface may locally exceed 135°C, the area is so small that there is not enough heating effect to cause ignition.

Objects larger than 10cm^2 will, if simple apparatus rules do not apply, require testing by the certification authority, but usually will end up with a T4 rating.

10 Inductance and Capacitance curves are shown in Appendix 2.

Section 3

This section is mainly aimed at the end user, and contains information on apparatus selection, installation, earthing, maintenance and training.

The Code of practice or other applicable publication for the country concerned should be consulted for additional information.

CHAPTER 14

Apparatus Selection and Procurement

Apparatus Selection and Procurement

Introduction

This chapter is mainly intended for those who have to specify and purchase apparatus made by other manufacturers.

Whilst attention is often given to those who design and install apparatus, the specifier and purchaser are frequently overlooked. It must be remembered that these functions have a major influence on the safety and overall integrity of the electrical system, and attention to detail at this stage is most important.

Much of the information in this chapter summarises more detailed coverage elsewhere in the book, and the reader should, if necessary, refer to those detailed sections.

Electrical Apparatus Selection

Electrical apparatus for hazardous area installation and use needs to be selected to be suitable for:

The hazard category
The temperature class
The zone of use

in addition to any environmental and operational requirements.

It is important that all these aspects are considered so that electrical apparatus conforms to the requirements for the severity of the hazardous area.

Sometimes there may be particular preferences for a project or plant: for example that a specific method of protection is used, or that a design conform to another existing plant or site or company standard. Whilst such criteria are perfectly understandable, it is worth remembering that new techniques are always emerging, and to be confined to a set of guidelines developed some time ago may prevent the most suitable approach now available being used. There can also be significant economic and safety penalties in not having an open approach to the selection and choice of apparatus and method of protection, and these aspects should be most carefully considered before becoming tied to specific guidelines which are unrealistic or unnecessarily restrictive.

There is no doubt that careful planning at the specification and procurement stage can pay dividends later on, and some clear plan as to the most suitable methods of protection and certification standards to be invoked should be evolved early in any project.

Among the aspects to be considered here are:

Is there a preferred Certification Authority or Standard?

Is there a preferred method of protection?

Is there a minimum acceptable level of environmental protection?

Although most Codes of practice state that no unnecessary electrical apparatus should be located in a hazardous area, it is important to view such statements in a realistic context, and there are often situations where additional apparatus may improve overall safety. For example if removing apparatus from the hazardous area means running a sampling pipe over a long distance just to locate a sensor in a non-hazardous area, the decision cannot be justified. (Such a situation will, in any case, probably extend or change the shape of the hazardous area.)

Overall good engineering practice should always be applied to the choice and location of apparatus, and there is little doubt that to run long electric cables to locate an instrument in a less severe zone or non-hazardous area, will be less safe than using a correctly installed intrinsically safe 'ia' instrument in a zone 1 or 2 area. There will probably be very little cost difference between the 'ia' installation (which is safe enough even for zone 0 use) and the alternative of running long leads and locating an instrument elsewhere. When the safety of the installation, including possible cable faults and deterioration is considered, siting intrinsically safe apparatus in the hazardous area will certainly be safer than the remote installation.

Thus, although it clearly makes good sense to minimise the amount of electrical apparatus in the hazardous area, this approach should not be taken to extremes, and the overall safety of the installation should always be considered.

Certification Authority

Usually, electrical apparatus will be certified or approved by some National Certifying Body or Test House. The more usual agencies are covered in Appendix 6, and examples of the marking and coding they use will be found in Appendix 1. In general, within Europe, the CENELEC Standards are now used by all certification bodies, but because some Test Houses are more diligent in their assessment than others, many end users, not unreasonably, have a preference for apparatus which has been certified by some particular authority: such as PTB or BASEEFA.

In theory at least, within the EEC all certification bodies should be equal, but, at the time of writing, this most certainly is not the case and it is doubtful if it ever will be, since technical Standards always require some interpretation and allow some latitude. Thus the Certification exercise always reflects the human aspect of the individual certifying officer, and the general attitude of a specific authority.

Certification Standards

The next stage is to establish which Standards are acceptable to the project for certification purposes. Manufacturers often sell apparatus which has been in production for many years. Such apparatus will probably have been certified to an old and perhaps obsolete Standard. Since a certificate, once issued, is usually valid for all time, a manufacturer may perfectly legitimately continue to sell apparatus even if the Standard to which the design was originally certified is no longer current. Indeed, many end users still prefer certified flameproof apparatus ('d') to the old BS 229 Standard, believing it to be more robust than its modern counterpart.

It is wise to state which Standards are acceptable. For example, intrinsically safe apparatus in the UK may be certified to the current CENELEC Standards, (EN 50 020 = BS 5501 Part 7) or to the BASEEFA document SFA 3012 which, although still used, has now largely been superseded by the CENELEC Standards. It may also have been certified to the original British Standard for Intrinsic Safety: BS 1259. Some companies and organisations have a policy to use only CENELEC Certified apparatus on a new project, and indeed, some EEC countries actually have internal regulations in force from their Factory Inspectorate which require EEC certified electrical apparatus to be used for new plant and projects. In the author's view this is a disappointing development, since it rules out the use of some excellent and perfectly safe apparatus, and has even prevented the selection of apparatus which would have in fact been the best for the task in hand. Thankfully, at least at the time of writing, no such pointless regulation exists in the UK.

Remember, National Standards give Ex... coded apparatus, whereas CENELEC certified apparatus gives EEx... coded apparatus.

In general, unless there are specific conditions or project requirements which dictate otherwise, apparatus certified to National Standards is really only suitable for use in the country in which it was certified. So a BASEEFA certified Ex... item will be acceptable in the UK, a PTB Ex.. item acceptable in Germany and so on.

Apparatus Suitability for Gas Group

Having established the acceptable Certifications and Standards, the apparatus selected must be suitable for the hazard category in which it is to be installed. This should have been specified in terms of the gas group, and the task here is to ensure that apparatus is at least safe enough for the gas group concerned.

With few exceptions, (notably Acetylene and Carbon Disulphide, where special care should be taken to ensure apparatus suitability) the following rule applies:

Group IIC apparatus is suitable for installation in Gas Group IIA or IIB or IIC

Group IIB apparatus is suitable for installation in Gas Group IIA or IIB, but **not IIC.**

Group IIA apparatus is only acceptable for installation in Gas Group IIA.

Apparatus marked only with II is suitable for all Gas Groups in Group II, i.e. IIA, and IIB, and IIC.

The appropriate code should be included on any specification, order, tender, certificate and, ultimately, on the apparatus label.

Apparatus Suitability for Temperature Class (T-Class)

The temperature classification of the apparatus (T-Class) must be specified, since, regardless of the gas group, apparatus needs to be safe in terms of its surface temperature. Again, this information should be available for the project, and the requirements must be followed. The T-Classification numbers T1 - T6 relate to the maximum surface temperature, and the higher the T-Class, the lower is the maximum surface temperature of the apparatus. So, if the project involves a hazard which has an auto ignition temperature of 230°C, T3 would be an appropriate temperature class, and apparatus marked T3, T4, T5, or T6 is acceptable. Beware of specifying a T-Class which is more restrictive than necessary however, since from the apparatus design point of view, keeping the surface temperature low is a feature of apparatus power, and can be quite restrictive. T6 apparatus is really only needed in carbon disulphide atmospheres, and there is no point in specifying T6 if the hazard is acetone. If the ignition temperature of the hazard is 230°C, then providing it is not subjected to a temperature in excess of 230°C, it will not ignite from hot surfaces. Greater safety will not be achieved by specifying a requirement for T4 or T5 or T6 apparatus.[1]

Selection of Method of Protection

As described in Chapter 3, some methods of protection are more suitable for some applications than others. There may also be a preference to keep to one or two methods of protection throughout a project to make installation and operation easier. Although such policies usually end up being broken for various 'special' items, it certainly often makes sense to minimise as far as practicable the different methods of protection used on any plant or project since this will assist subsequent installation and maintenance activity. However, since some methods of protection are suitable only for zone 2 use, and some for only zones 1 and 2, the selection and preference of method of protection needs to be given careful thought.

One aspect which certainly is worthy of consideration, is a general guideline to use a specific method of protection wherever possible for a particular application. For example:

Instrumentation: Intrinsic Safety
Junction boxes: Increased safety

Almost all instrumentation functions can now be achieved using intrinsic safety techniques, and the advantages of installation, cabling and maintenance costs usually make this an attractive as well as cost effective approach, regardless of the zone of installation. As far as junction boxes are concerned, it is virtually certain that no junction boxes will be required in a zone 0 area, and so providing they are suitable for zones 1 and 2, the zone selection criteria is satisfied. Standardizing on Increased Safety 'e' gives a solution which is suitable for both zones 1 and 2, and such junction boxes will probably be suitable for all circuits. Indeed, especially if purchasing a large number of such enclosures, they might as well

210

be specified even for the intrinsically safe circuits, since, although not required in such circuits, it allows a common type of enclosure to be used throughout the plant. [2]

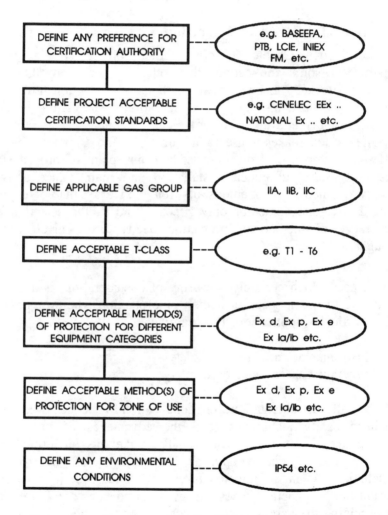

Figure 14.1 *Selection Procedure for Methods of Protection*

Figure 14.1 shows the basic steps in successfully specifying the method of protection to be used in hazardous areas.

Use of Apparatus which is Not Certified

There may be occasional situations when apparatus which is not in any way certified or approved is required. There should never be any situation where uncertified apparatus is allowed in either zone 1 or zone 0,[3] but it may be possible to accept uncertified apparatus for zone 2 use only, under certain circumstances.

In all cases expert advice should be sought to ensure that uncertified apparatus will be acceptable to the project, and a clear company policy on the use of uncertified apparatus in zone 2 should be evolved to avoid costly and time consuming errors being made at the design and specification stage.

Apparatus which 'Conforms to ..'

When checking quotations and tenders, or placing orders for apparatus, beware of wording such as:

"Generally conforms to the requirements of flameproof protection."

or "Manufactured in accordance with CENELEC Standard"

or "Conforms to .... "

Such terms do not equate to certification and should not be accepted as such. Unless it is intended to use uncertified apparatus for zone 2 installations, the type of certificate, certificate number and certification code (for example EEx d IIB T4) should be requested from the manufacturer at quotation stage, and any order should require a copy of the certificate to accompany the goods delivered.

Claims such as those above which could mislead the end user are often made about motors for zone 2 use. Although there are, of course, many reputable manufacturers who have either carried out a very full assessment themselves, or obtained certification from BASEEFA for an Ex n motor to BS 5000: Part 16, there are a number of motors on the market which have serious omissions in their compliance with such Standards. The best advice is to study the documentation carefully, question the supplier if the position is not clear, and, when the apparatus arrives, check the label and accompanying documentation for certification information.

Environmental Protection

When selecting apparatus, remember that protection for use in a hazardous atmosphere does not necessarily mean that apparatus is suitable for outdoor conditions. If apparatus is to be located outside, then some protection against rain, dust, sand etc. will be needed. There are two methods of specifying environmental protection; the IP code[4] and the NEMA code[4] which is more often used in North America. As a general guide, IP 54 is suitable for general outside use, and IP65 will give good protection against all but the very worst conditions. It should be noted that many manufacturers claim an IP protection level even if the apparatus has not been fully tested. The tests for IP ratings are quite stringent and apparatus which may appear to be well protected often fails the test to the full specification. If environmental protection is important, it is worthwhile establishing from the manufacturer if full type tests have been carried out by an independent body to confirm the IP rating.

Also remember that corrosion problems may lead to a requirement to avoid aluminium, etc.

Environmental protection is especially important where the method of protection is flameproof 'd' or explosionproof. The concept of flameproof protection often means that there is a gap around joints and flanges in the equipment.[5] Unless gaskets have been included as part of the apparatus design and certification, they cannot be added subsequently by the end user without the risk of invalidating the certification.

Check List for Apparatus Specification and Procurement

1 Does the apparatus have to be located in the hazardous area?

2 Check the hazardous area details; Gas Group, T Class and Zone.

3 Note any particular environmental requirements, high (or low) ambient temperature, IP rating etc.

4 Note any particular problems of corrosion which may be experienced.

5 Define any preference for materials, eg stainless steel or plastics etc.

6 Define any preference for Certification Authority.

7 Define any preference for Certification Code (Ex or EEx)

8 Define preference(s) for method(s) of protection.

9 Define requirements for gaskets etc. (Especially on flameproof protection, where gaskets must be included by the manufacturer to certification conditions.)

10 Check manufacturer's information, in particular, is there an 'X' of 'B' condition on the Certificate, and is the apparatus actually certified.

11 Request copy of Certificate with goods. (A copy should be retained for the record system to enable checks to be made later, when, perhaps, the apparatus label has become unreadable.)

12 Ask the manufacturer if there are any particular installation requirements or advice, and check that the apparatus can be installed to the relevant Code of practice or requirements of the country of use.

214

Notes and References

1   Clearly there are limits to this approach. It should be noted that published auto ignition temperatures can vary by a few degrees from publication to publication, and it would be unwise to unnecessarily cut any corners. For example, if the hazard has an auto ignition temperature of 301°C, then, although T2 will in theory be acceptable, it might be prudent to use T3 rated apparatus.

As an alternative, it may be possible to obtain from the manufacturer a copy of the Certification Test Report. This document may indicate the actual temperature measured which led to the T Class which has been awarded. For example, if the certification work established a maximum surface temperature of 240°C, then the apparatus would be awarded a T2 rating. Clearly however, such apparatus would be acceptable for the above hazard.

In practice, borderline cases such as these are rare, and the T Class information is usually sufficient for apparatus selection criteria.

2   Careful attention must be given to the correct labelling of junction boxes which contain intrinsically safe circuits. Refer to Chapter 15.

3   It may be acceptable to use uncertified apparatus in zone 0 or zone 1 in certain circumstances such as research, development work and pilot plant. In such situations the user should obtain a 'Document of Conformity' from a competent third party to show that the apparatus has been examined and tested and is considered suitable for the application.

4   A full table of Ingress Protection (IP) and NEMA codes is given in Appendix 7.

5   Refer to Chapter 9 for more information on flameproof enclosures.

CHAPTER 15

Installation

Installation

STANDARDS AND REFERENCE DOCUMENTS	
BS 5345	Code of practice for selection, installation and maintenance of electrical apparatus for use in potentially explosive atmospheres. (Part 1 and other parts depending on the method(s) of protection involved)
IEC 79-14	Electrical installations in explosive gas atmospheres
ANSI/ISA-RP 12.6	Installation of Intrinsically Safe Systems for Hazardous (Classified) Locations (USA)
ANSI/ISA-RP 12.1	Electrical Instruments in Hazardous Atmospheres (USA)

Although much emphasis is placed on the correct selection of electrical apparatus for hazardous areas, its correct installation is of equal importance. Furthermore, the installation is usually in the hands of the end user, and thus is an area which is not affected to any great extent by the apparatus manufacturer, apart from possible installation data and advice.

There is no doubt that many end users believe that the whole business of electrical safety in hazardous areas begins and ends with the selection of apparatus. Nothing could be further from the truth. The correct installation is of paramount importance to the overall safety, and needs to be given the most careful consideration.

The relevant Code of practice should always be consulted for specific details and requirements or recommendations for each method of protection.

The main aspects which will receive attention is this chapter are:

Cables and wiring
Cable glands
Terminals and terminations
Junction boxes

Cables and Wiring

Cables and wiring in hazardous areas need to provide a level of protection suitable for the zone of use. Thus the requirements and recommendations of the various codes are based on the following principles:

Zone 0   The only method of protection permitted is intrinsic safety ('ia') and thus the cables must be suitable for intrinsically safe circuits. [1]

Zone 1   The cables must be suitably mechanically protected to provide a reasonable level of assurance that normal cable damage and wear and tear will not cause a sparking condition.

Zone 2   Cables should still be mechanically protected. Any cable suitable for zone 1 is acceptable, and in addition other cables with less protection, including unarmoured industrial PVC cable may be used where the risk of mechanical damage is shown to be slight.

In the main, cables apart from intrinsically safe circuits require reasonable mechanical protection, and this group will be examined first.

Clearly there is little point in protecting the electrical apparatus (for example by using flameproof or increased safety protection) if the interconnecting cable can give rise to an incendive condition. Thus the cables, their method of installation, supports and terminations need to achieve an overall level of security against a possible arcing or sparking condition.

Although there are minor differences in the recommendations for different countries and certain specific industries, the general requirements are very similar and the UK Code of practice BS 5345 is a good representative guide.

BS 5345: Part 1, recommends the following cables as suitable for use in zones 1 and 2.

a) Cables drawn into conduit systems

b) Cables that are otherwise suitably protected against mechanical damage.

Examples [2] of cables in (b) are:

- Thermoplastics or elastomer insulated screened or armoured cable with or without lead sheath and with overall sheath of polyvinylchloride (PVC), chlorosulphonated polyethylene (CSP), polychloropropene (PCP), chlorinated polyethylene (CPE), or similar.

- Cables enclosed in seamless aluminium sheath with or without armour, with an outer protective sheath.

- Mineral insulated metal sheath cable.

- Thermoplastics or elastomer insulated flexible cable or cord with a flexible metallic screen or armour and PVC, CSP, PCP, CPE or similar overall sheath.

- Thermoplastics insulated cable with semi-rigid sheath.

All these cables provide a reasonable level of mechanical protection, so as to minimize the risk of cable damage or breakage which could create an incendive situation. See Figure 15.1.

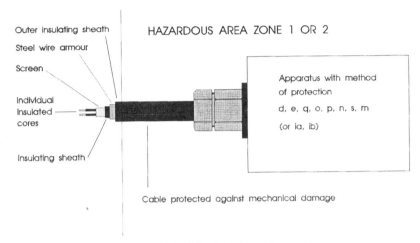

Figure 15.1   *Protection of Cable in Hazardous Areas*

Where mineral insulated metal sheathed cable (MIC) is used, particular care should be taken to ensure that insulation properties are not degraded by dampness. The correct termination device must be used.

Conduit Systems

Where conduit systems are used, particular care should be taken to ensure that correctly installed stopper glands or stopper boxes are used at the entry point to any apparatus. This is especially important with flameproof ('d') apparatus.

Conduit systems are not recommended for outdoor locations, and aluminium, plastics and flexible conduits should not be used in zone 1 or zone 0 locations. In addition, any corrosive action which may degrade the conduit should be carefully considered, and may make the use of the conduit undesirable.

220

Conduit systems should be carefully installed to ensure that there are no sharp edges or protrusions which could cause damage when the cable is being pulled.

It should also be remembered that conduit and cable ducts may act as a pipe and transfer the hazard from one place to another. It is thus important to ensure that any conduit is sealed or stopped at the entry point to the non-hazardous area, as well as at the field termination end.

For similar reasons, any holes in walls or other solid divisions between the hazardous and non-hazardous area where cables or ducts pass, should be blocked and sealed as far as possible. Sometimes a sand trap is suitable for this purpose. Figure 15.2.

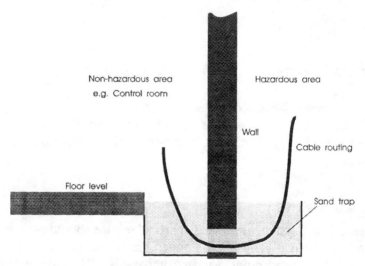

Figure 15.2   *Sand Trap for Cable Routing at Junction Between Hazardous and Non-Hazardous Area*

Cables for Intrinsically Safe Installations

For cables used in intrinsically safe installations, the problems of mechanical damage are of less concern since even if the cable is damaged, the result should not be ignition capable. Thus, although intrinsically safe circuits may use screened and/or steel wire armoured cable, (SWA) there is no requirement for this. Figure 15.3.

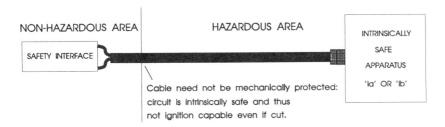

NON-HAZARDOUS AREA | HAZARDOUS AREA

SAFETY INTERFACE

INTRINSICALLY SAFE APPARATUS 'Ia' OR 'Ib'

Cable need not be mechanically protected: circuit is intrinsically safe and thus not ignition capable even if cut.

Figure 15.3 *Cable With Intrinsically Safe Circuits*

Where intrinsically safe circuits use screens and metallic sheaths, the normal earthing arrangements are to earth the screen at one point only; usually the non-hazardous area end (normally to the intrinsically safe interface earth connection), and to earth the sheath (SWA) at each cable gland. (Figure 15.4) This is discussed further in Chapter 16.

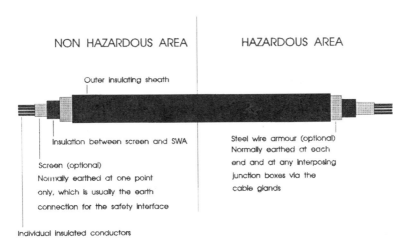

NON HAZARDOUS AREA | HAZARDOUS AREA

Outer insulating sheath

Insulation between screen and SWA

Screen (optional)
Normally earthed at one point only, which is usually the earth connection for the safety interface

Steel wire armour (optional)
Normally earthed at each end and at any interposing junction boxes via the cable glands

Individual insulated conductors

Figure 15.4 *Earthing of Screens & Sheaths in Intrinsically Safe Circuits*

The conductor sizes of cable need to be sufficient to ensure that the cable temperatures do not exceed the T-Class of hazard. Providing a temperature class better than T4 (that is T5 or T6) is not required, the following table may be used to establish suitable conductor sizes.

Maximum Current (A)	1.0	1.65	3.3	5.0	6.6	8.3
Cross sectional area of conductor in mm^2	0.017	0.03	0.09	0.19	0.28	0.44

Table 15.1 *Conductor Sizes for Various Currents to Achieve T4 Rating on Interconnecting Cables*

The maximum current should be regarded as the maximum current under fault conditions. This is normally easily ascertained by using the maximum output parameters of the safety interface unit. In practice, the types of cables normally used in intrinsically safe systems will be perfectly acceptable for most intrinsically safe circuits. Only where an intrinsically safe circuit uses unusually large currents is there likely to be any problem. (It should also be remembered that if currents are large on intrinsically safe circuits, there may well be a problem with cable inductance, and the cable parameters specified on the system certificate and system drawing should be checked carefully.)

Although there is no requirement to do so, it is common practice for intrinsically safe circuits to be run in cables which have a blue outer insulating sheath. The colour blue is the colour used to indicate intrinsic safety, and apart from other considerations, using blue cable for the IS circuits greatly assists in identifying cables, junction boxes and IS apparatus which may be worked on live.

To avoid confusion, blue covered cables should not be used for circuits which are not intrinsically safe.

The use of some blue cables and others which are not blue for IS circuits should also be avoided since this too can easily lead to confusion.

Use of Multicore Cables

Multicore cables may be used for combinations of different intrinsically safe circuits, but on no account should intrinsically safe and non-intrinsically safe circuits be mixed in the same multicore.

Installation Requirements for Cables in Intrinsically Safe Circuits

The cable should be able to withstand a 500 volt insulation test between different cores, between cores and any screen, and between different screens, and between screens and any armour or metallic covering. Normally this test should be carried out at installation stage on the installed cable since it assists in indicating if the cable has sustained any damage during installation. A 'Megger' will give a good practical test, although the correct test method is to gradually increase the voltage to 500V ac rms, and to hold the voltage at 500V for one minute. There should be no breakdown.

Cables for Zone 2 Installations

The cable requirements indicated earlier in this section as recommended by BS 5345: Part 1 may be relaxed somewhat for zone 2 installations, to include any stranded, armoured or unarmoured industrial wiring where the conductors are insulated throughout their length and adequately protected against damage. This protection against damage, where not provided by SWA or similar, can be achieved by trunking or conduit, and, if it can be shown that the risk of mechanical damage is small, then no additional protection at all is required. In practice, most end users require some form of mechanical protection for cables anyway, and so apart from some intrinsically safe circuit loops it is rare to use unprotected cables.

224

Consideration of Environmental and Operational Conditions

Cables are frequently damaged by poor installation practices and procedures such as insufficient clamping or excessively tight bends, or routing over sharp edges. Cable clamping is especially important on long vertical routes, since the weight of cable can cause premature ageing and stress. Remember also, that cables specified in the Code of practice as being suitable for hazardous area use are not necessarily suitable for wet or corrosive conditions.

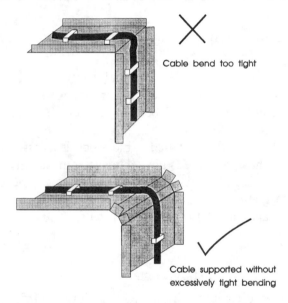

Cable bend too tight

Cable supported without
excessively tight bending

Figure 15.5   *Examples of Correct and Incorrect Cable Tray Bends.*

Where flexible leads are likely to be subjected to excessive vibration, special consideration of the type of cable should be given, and enhanced inspection at critical wear points may be necessary.

Care should be taken to ensure that cable insulating material will not be adversely affected by fumes or deposits of any chemicals likely to be present. Again, additional inspections and/or expert advice may be necessary to ensure that cables are kept in good order on plants where such chemical attack is likely.

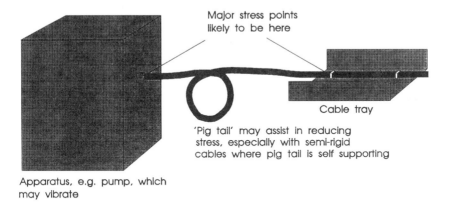

Major stress points likely to be here

Cable tray

'Pig tail' may assist in reducing stress, especially with semi-rigid cables where pig tail is self supporting

Apparatus, e.g. pump, which may vibrate

Figure 15.6   *Cable Considerations for Items with Excessive Vibration.*

Cables using Light Alloy Coverings (e.g. Aluminium)

Cables (and equipment) with light alloy coverings are probably best avoided in most situations, because they tend to attract rapid corrosion by the build up of oxides e.g. Al_2O_3. In addition to possible corrosion problems, it is necessary to consider the increased possibility of frictional sparks from light alloy metals. Such metals should, in any case, not be used in zone 0 areas. For zone 1 installations, if light alloy metals must be used, care should be taken to ensure that the likelihood of frictional sparks is minimised, for example by only using such metals where the risk of impact is minimal, or if such cables have a heavy and robust outer protective covering.

For zone 2 installations, the precautions indicated for zone 1 should still be considered, especially if there is risk of high energy impact or if the impact may also lead to a release of flammable material.

Cable Glands [3]

Cable glands often provide the access means for connection of electrical apparatus. Thus the cable glands need to ensure that the methods of protection of the apparatus which they enter is maintained. To begin with, consider the main criteria of the different methods of protection. (Table 15.2)

METHOD OF PROTECTION	CONSIDERATION FOR CABLE ENTRY POINTS
Flameproof	Apparatus must withstand internal explosion without transmitting it to outside of enclosure
Pressurized	Apparatus must provide IP54 protection, and meet any pressure test requirements
Increased Safety	Apparatus must provide IP54 protection and withstand impact test
Oil Immersion	Apparatus must provide IP54 protection, and meet any pressure test requirements
Sand Filling	Apparatus must provide IP54, and meet any pressure test requirements
N-Type	Must provide IP54 (IP30 indoors)
Intrinsic Safety	Must provide IP54, unless associated safe area apparatus
Special Protection	Normally IP54, plus any additional requirements
Encapsulation	Cable normally encapsulated into apparatus as an integral lead, thus cable glands not normally applicable to end user

Table 15.2   *Cable Entry Considerations for Different Methods of Protection*

NOTE: In addition to maintaining the protection required, the cable gland also provides a means of anchoring the cable so that conductor terminations are not stressed if the cable is pulled. This aspect is important with all methods of protection, but perhaps especially with 'e' and 'n', where a poorly anchored cable, if pulled, could give rise to a sparking condition.

Cable Glands: Flameproof Protection

Clearly, where the apparatus is flameproof, the cable gland needs to maintain that protection integrity, and an Ex d or EEx d component certified gland must be used. (Figure 15.7)

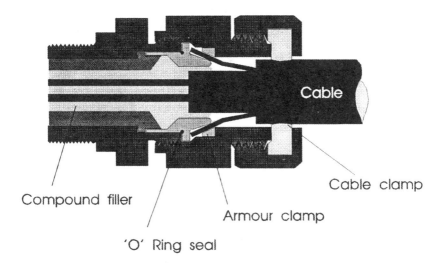

Figure 15.7   *Example of Ex d Cable Gland*

The use of the correct gland is of course negated if the cable is incorrect for the gland, or if the cable is not correctly made off in the gland, or if the gland thread is not correct to the thread in the apparatus. Normally the flameproof enclosure will have tapped cable entry holes provided, and the thread will be present through the thickness of the wall of the material of the enclosure. The end user should avoid drilling or tapping additional cable entry holes unless it is absolutely unavoidable, in which case the utmost care should be taken, and the grade and quality of both drill and tap checked before proceeding. The manufacturer of the apparatus should be contacted and advised of the intention to drill additional holes, since he may have specific requirements to keep the certification valid, (for example on the location or size of such holes) or may insist that the apparatus is returned for this operation. Alternatively, some manufacturers will allow such work to be

228

carried out on site if it is under the supervision of one of their own engineers. Very often, and especially for such apparatus as motors and luminaries, a number of different cable entry gland plates are available from the manufacturer, and it may be possible to purchase a different plate which allows the required cable configuration.

Unless required for environmental protection, and permitted by the gland certification, sealing washers need not be used. Where they are used, the minimum thread engagement must still be achieved.

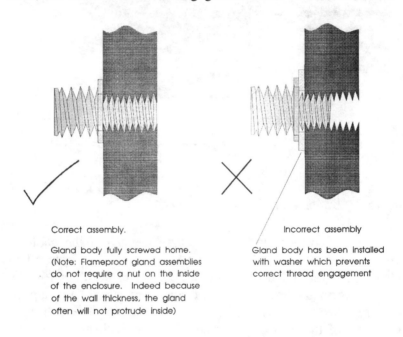

Correct assembly.

Gland body fully screwed home.
(Note: Flameproof gland assemblies
do not require a nut on the inside
of the enclosure. Indeed because
of the wall thickness, the gland
often will not protrude inside)

Incorrect assembly

Gland body has been installed
with washer which prevents
correct thread engagement

Figure 15.8 *Installation of Ex d and EEx d Cable Glands*

Cable Glands for Other Methods of Protection

As will be seen from Table 15.2, for most methods of protection apart from 'd', the main requirement is for an IP54 level of protection. Although certified glands, for example increased safety 'e', are available, most industrial cable glands which are capable of providing an IP54 seal are suitable. Again, not only is the gland itself important, but the cable to gland, and gland to enclosure aspects must also be considered, since the IP54 integrity can easily be lost at these points. A sealing washer is a normal requirement for these methods of protection.

NOTE: Some users choose to standardise on flameproof glands for all methods of protection throughout an installation to avoid the possibility of incorrect glands being used on flameproof apparatus. It should be remembered that if such glands are issued from stores with a sealing washer, the washer must not be used on flameproof applications. Equally, if no sealing washer is available, then it must be obtained before using a flameproof gland on a non-flameproof installation. In general, this practice of standardising on flameproof glands causes a great deal of confusion and is best avoided.

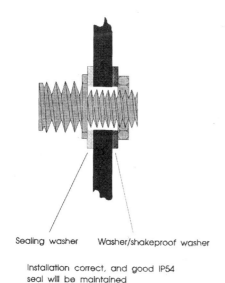

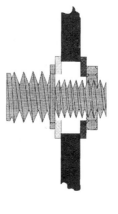

Sealing washer   Washer/shakeproof washer

Installation correct, and good IP54
seal will be maintained

Incorrect assembly

The hole in the glandplate is too large
and the gland is only being held tight
by the sealing washer. This will lead
to degraded IP rating in due course.

Figure 15.9   *Correct and Incorrect Assembly of Gland to a Glandplate (Protection Methods other than Flameproof)*

For some situations, for example oil filled apparatus EEx o, it may be necessary to use cable glands which will, when correctly installed, withstand particular pressure test requirements. These can usually be achieved by a flameproof gland which incorporates suitable internal sealing compound, installed with some sealant around the thread. The manufacturer of the apparatus (certificate holder) should be consulted in such cases to avoid using incorrect glands and invalidating the method of protection.

It is important to realise that some cables exhibit a property known as cold flow. This property is best explained by understanding that some plastic materials actually 'flow' away from a pressure point. As far as cables are concerned, this means that a correctly tightened gland may become loose again after a few months, because the thickness of insulation has actually diminished at the clamping point. Cable manufacturers should be consulted to ascertain cold flow characteristics of cables used, and, if appropriate, special attention should be given to the tightness of cable glands at inspection stages. (see Chapter 17)

If apparatus has been removed for repair, or if cabling has been removed, cable glands probably need to be replaced with new ones, because the compression parts of a gland may loose their elasticity and not correctly seal the new/renewed cable entry. Again, this aspect is of particular importance with cable glands to flameproof apparatus.

Terminals and Terminations

The electrical terminations on apparatus in hazardous areas are of crucial importance since this is the main part which the installer and end user can affect. Examples of problems which can occur are:

Incorrect conductor size for termination devices.

Loose ferrules or spades.

Loose clamping nuts or screws.

Shake proof / star washers not fitted.

Stray conductor strands invalidating creepage and clearance distances and leading to possible arcing.

Although conductors should, of course, be correctly terminated for all methods of protection, it is extremely important for increased safety 'e', N-Type protection 'n', and intrinsic safety 'ia'/'ib' where degradation may give rise to an unprotected ignition state. It should be remembered that types of protection 'n' and 'e' rely for their safety on the apparatus being non-sparking. If arcs or sparks do occur, there is no additional protection of, for example, a flameproof enclosure, to protect against ignition of the hazardous area. So it is critical with these methods of protection to ensure that terminals and terminations are correctly secured and tightened, if necessary to the torque recommended by the terminal manufacturer. For example, a common type of rail mounted terminal for 4mm² stranded conductors has a recommended tightening torque of 0.3 to 0.5 Nm to ensure correct operation of the vibration proof qualities.

If the conductors are aluminium, then it is especially important to re-check the tightness about one month after installation in case distortion of the conductor has resulted in a loose connection. (The conductor should be cleaned prior to insertion, and a smear of Vaseline or petroleum jelly smeared onto the conductor immediately after cleaning to prevent rapid oxidation).

Where stranded conductors are used, care should be taken to ensure that all the strands are secured, that individual strands are not broken and missing, and that there are no stray conductors which could short to adjacent conducting parts. Normally, stranded conductors should have a ferrule crimped to the conductors. The crimp must be secure, correctly clamping all the conductor strands and shielding the stripped insulation.

The dangers of loose strands causing sparking conditions is worthy of particular attention for methods of protection 'e', 'n', and, where multiple circuits share a common junction box, intrinsic safety.

232

Normally, each termination should be identified with a terminal number and a cable ident sleeve so that correct connection can easily be checked against loop diagrams or installation drawings

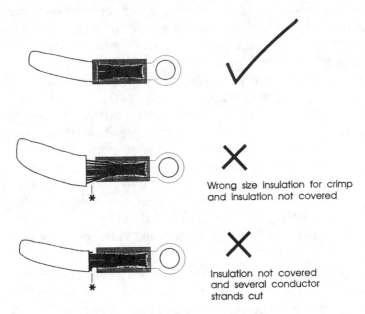

Wrong size insulation for crimp and insulation not covered

Insulation not covered and several conductor strands cut

Figure 15.10   *Crimping of Stranded Conductors*

Junction Boxes

Junction boxes normally involve one of the following methods of protection:

Flameproof	'd' : suitable for zones	1 and 2
N-Type	'n' :	2 only
Increased safety	'e' :	1 and 2
Intrinsic Safety	'ib':	1 and 2
Intrinsic Safety	'ia' :	0, 1, and 2 *

*   It is extremely unlikely that any junction box will actually be required to be located in a zone 0 area. The grade of intrinsic safety 'ia' or 'ib' of a junction box in an intrinsically safe

circuit, is purely a function of the grade of intrinsic safety of the circuits contained in the junction box.

Very often, it may be feasible to standardise on one type of junction box throughout an project, plant, or installation. For example if increased safety 'e' junction boxes are selected, they are suitable for both zones 1 and 2, and, although not needed for zone 2 applications, or where all the circuits in the box are intrinsically safe, (when the junction box may be treated as simple apparatus, see Chapter 13) an increased safety junction box does provide a degree of protection of IP54, and would be perfectly suitable for intrinsically safe circuits. If the box does contain intrinsically safe circuits, a warning label should be attached to the box cover to indicate that intrinsically safe circuits are present.

TAKE CARE

THIS ENCLOSURE

CONTAINS

INTRINSICALLY SAFE

CIRCUITS

Figure 15.11   *Label for Junction Boxes containing Intrinsically Safe Circuits*

Ideally, intrinsically safe and non-intrinsically safe circuits should not be mixed within a junction box. If this is not possible, then the warning notice must make it clear that both intrinsically safe and non-intrinsically safe circuits are present, since this will mean that the enclosure cannot be opened for live maintenance as would be permitted if only intrinsically safe circuits were present.

It is also important to ensure that the correct segregation between intrinsically safe and non-intrinsically safe circuits is provided. This can best be achieved by the introduction of some physical barrier, for example an insulating shield between intrinsically safe and non-intrinsically safe circuits. If such a shield is not possible, then a distance of at least 50 mm should be preserved between live conducting parts of intrinsically safe and other circuits.

The choice of material for junction boxes is mainly determined by the environmental conditions. For example in outside positions, plastics may be preferable to steel. Even so, some plastics may deteriorate if exposed to direct sunlight for long periods, and in such situations, stainless steel might be the best choice.

Isolation of Electrical Apparatus

Apart from the technique of intrinsic safety, electrical apparatus in the hazardous area should not be opened whilst the circuits are live.[4] Thus there needs to be an effective means of isolation for each circuit, and this isolation device should be located outside the hazardous area so that the whole of the circuit in the hazardous area can be de-activated.

The isolation device should have the facility for locking in the 'off' position, or for the removal of fuses or some other means of ensuring that an isolated circuit is not inadvertently made live whilst the method of protection of items of apparatus in the circuit has been violated, for example during maintenance or inspection work.

If there are additional isolation devices located within the hazardous area, then this apparatus clearly needs to have some suitable method of protection. (Normally a flameproof 'd' switch).

Some electrical apparatus contains more than one circuit, and it may be that for example a motor has vibration monitoring circuits or temperature sensing elements which are still live after the main power has been removed. In such cases the apparatus should bear a label stating that this condition applies. The label is normally affixed by the apparatus manufacturer as part of the certification requirement. Apparatus bearing such warnings should be treated with special care, and personnel should be aware of the continued danger of electrical ignition if the respective sensing circuits have not also been isolated. Unless the sensing circuits are themselves intrinsically safe, then if they are to be left in a live state whilst the covers of the main apparatus are removed, a hot work permit will be needed anyway to cover the safety of this aspect.

Combinations of Apparatus with Different Methods of Protection

With the exception of intrinsic safety which is a system concept, all other methods of protection may be mixed within an installation and a circuit loop. For example it is perfectly possible to have a flameproof motor (located in say zone 1) with the supply cabling terminated in an increased safety junction box (also in zone 1). The method of protection is purely a function of the apparatus itself, and (providing it is not electrically or environmentally operating outside its constraining conditions) is not affected by other electrical apparatus elsewhere in the circuit.

Although it is usually both impractical and impossible to standardize on just one or two methods of protection, it is often possible to minimise the methods of protection used for the majority of items. This aspect is discussed further in Chapter 14.

Identification of Apparatus

It is most important that apparatus is correctly identified, both with the manufacturer's certification information, and with the user's additional site information of loop number, circuit reference or unique item code.

Unfortunately, the requirements for certification labelling are not especially clear in the Standards, and often the kind of labels preferred by the certification authorities are not at all suitable for the duty expected of them when installed in the hazardous area. It is very common to find that, after a few years, the majority of the certification labels are no longer visible, or have fallen off, and the presence of the end user's identification plate is the only means of tracing exactly what is present and which method(s) of protection and certification conditions apply to the apparatus. In spite of the great improvement in industrial adhesives in recent years, experience still indicates that labels which are stuck on tend to fall off. Engraved plastic labels are probably best, but of course care needs to be taken not to invalidate the method of protection by drilling fixing holes! Many end users prefer to identify the apparatus by affixing such a label with wire around the cable entry point or adjacent to the apparatus. This has the advantage that identification may easily be cross checked if the apparatus has been removed for repair, but can lead to problems of identification if changes are made to plant layout or if wiring is renewed.

It is well worth giving the aspect of labelling careful thought, and evolving a system which is safe, suitable for the particular plant, and will provide continued identification of the apparatus.

Notes and References

1 Occasionally, special protection 's' may be suitable for zone 0 use. (See Chapter 11) In such cases the manufacturer of the apparatus should be consulted and the certificate examined for any particular cable requirements or installation conditions.

2 Some useful British Standards associated with cables are as follows:

BS 31   Specification. Steel conduit and fittings for electrical wiring. (1988 edition)

BS 5345   Code of practice selection, installation and maintenance of electrical apparatus for use in potentially explosive atmospheres (other than mining applications and explosive processing or manufacture)

BS 5308   Instrumentation cables

           Part 1   Specification for polyethylene insulated cables

           Part 2   Specification for PVC insulated cables

BS 731   Flexible steel conduit for cable protection and flexible steel tubing to enclose flexible drives.

           Part 1   Flexible steel conduit and adapters for the protection of electric cable

BS 4568   Specification for steel conduit and fittings with metric thread of ISO form for electrical installations

           Part 1   Steel conduit, bends and couplers

           Part 2   Fittings and components

BS 5467   Specification for cables with thermosetting insulation for electricity supply for rated voltages of up to and including 600/1000V and up to and including 1900/3000V.

BS 6004   Specification for PVC-insulated cables (non-armoured) for electric power and lighting

BS 6207   Specification for mineral-insulated copper sheathed cables with copper conductors

BS 6346   Specification for PVC-insulated cables for electricity supply

BS 6500   Specification for insulated flexible cords and cables

BS 6724   Specification for armoured cables for electricity supply having thermosetting insulation with low emission of smoke and corrosive gases when affected by fire

3   The following British Standards deal in more detail with cable glands. In the main they are concerned with the construction of glands, and thus may not be of particular interest to the end user.

BS 4121 *   Specification for mechanical cable glands for rubber and plastics insulated cable.

BS 6081   Specification for terminations for mineral insulated cables.

BS 6121   Specification for mechanical cable glands for elastomer and plastics insulated cables.

* Withdrawn

4   Live maintenance is permitted in certain circumstances if the area has been declared gas free, and a 'hot work permit' has been issued. Refer to Chapter 17 for more details.

CHAPTER 16

Earthing

Earthing

The subject of earths and earthing of electrical apparatus in hazardous areas always seems to cause disproportional problems and excitement. The specific requirements may vary from country to country, and the relevant Code of practice should always be consulted. This chapter sets out to explain the basic requirements and indicate the essentials for different methods of protection.[1][2][3]

Because the method of protection intrinsic safety ('ia' or 'ib') allows freedom on the types of cable which may be used for the interconnection of apparatus, there are some differences between the recommendations for this protection concept and other methods. In practice, however, the earthing requirements for intrinsic safety are often identical to those for other methods of protection, and indeed, often translate into normal engineering practice.

Earthing Requirements for Intrinsic Safety

As discussed in Chapter 13, intrinsic safety requires the use of some safety interface device, located in the non-hazardous (safe) area. First of all, consider that safety interface device to be a conventional zener safety barrier.

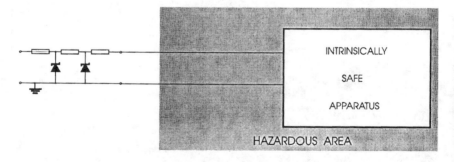

Figure 16.1 *Simple Intrinsically Safe Circuit with Zener Barrier*

As will be seen from Figure 16.1, one side of the barrier circuit is earthed. That is, the intrinsically safe circuit will, if the barrier is correctly installed, be earthed at one point; in the non-hazardous area. If the circuit were to be earthed in the hazardous area as well (Figure 16.2), then it would be possible for circulating earth currents in the cable to be in excess of intrinsic safety levels. With intrinsic safety, there is no requirement to use armoured cables which are protected against the possibility of damage or breakage, and thus the possibility of this conductor being broken needs to be considered. If excessive circulating currents were present when the conductor parted, then an incendive condition could result.

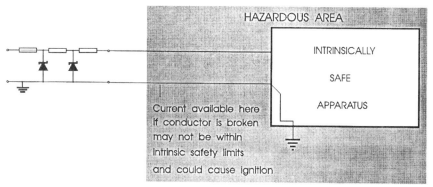

Figure 16.2 *Showing Circulating Currents if Intrinsically Safe Circuit is Earthed at Two Points*

Thus the normal requirement for intrinsic safety is that the circuit be earthed at one point only, this point normally being the earth connection for the intrinsically safe circuit in the safe area. The electrical circuit in the hazardous area should maintain a 500V insulation test to local earth or frame of the field apparatus. This last statement does not imply that the field apparatus itself should not be earthed locally, and indeed the act of mounting the apparatus normally achieves this, at least to some extent, if the case of the apparatus is metallic.

It is normal practice to locally earth the case of hazardous area apparatus to the plant structure. If an external earth stud is

242

provided, this should be used for the purpose, and the connection achieved should be such that it will not vibrate loose. Figure 16.3 summarises this position.

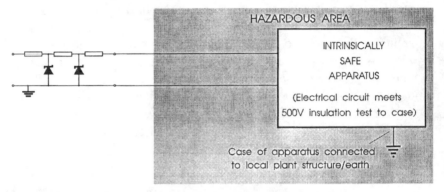

Figure 16.3   *Earth Connections of Hazardous Area Apparatus*

If the intrinsically safe circuit is such that the 500V insulation cannot be achieved, then there will normally be special instructions on the system documentation supplied by the manufacturer or on the System Certificate. A quite common position in this situation is to use a bonding conductor between the case of the field apparatus and the zener barrier earth busbar. (Figure 16.4) This bonding conductor will carry any circulating earth currents, and the conductor is substantial (at least 4 mm^2 and often 6 mm^2) so that it is unlikely to become broken with a sparking result. The system documentation should be studied to check if there is a maximum length imposed between the interface unit and the intrinsically safe apparatus.

Where galvanic isolators are used rather than zener barriers, there is no safety earth at the interface point, and thus even if there is a connection to earth in the hazardous area, incendive currents should not result. Thus an alternative approach where the 500V insulation test to earth is not attainable, is to use galvanic isolation interfaces.

If the field apparatus is uncertified 'simple apparatus', then it is the responsibility of the manufacturer when claiming compliance with the simple apparatus rules to ensure that the installation conditions

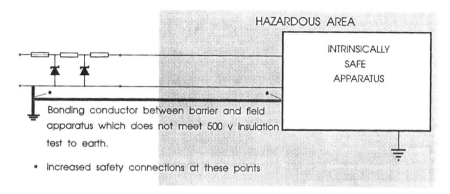

Figure 16.4   *Use of Bonding Conductor Where 500 Volt Insulation Requirement Cannot be Achieved*

can be met. Thus the manufacturer should be able to confirm that such a device will meet the 500V test requirements.

So far, the assumption has been made that the interconnecting cable will not be mechanically protected by a conducting armour. Although not required for intrinsic safety, many users opt to use steel wire armour cable throughout their installations, and the question then arises where and how should the armour be earthed? This question has been the subject of much discussion over recent years, since if the armour is earthed at both ends, this seems to counteract the 'earth at one point' requirement. However, if an armoured cable is used, it is unlikely to become broken, in which case the possibility of incendive sparks from circulating earth currents does not arise! If the user wishes to earth the steel wire armour at only one point, then it is normal that it will be connected to earth via the cable gland as it enters the field apparatus. (Assuming that the apparatus has a metallic case which is locally earthed to the plant structure.) It is, however, more normal practice to connect the armour to earth at both ends, and at any point where there are interposing cable glands. (For example at junction boxes.)

Note in Figure 16.5 that, at the non-hazardous area end, the armour is indicated as connected to the cabinet earth, rather than the barrier

244

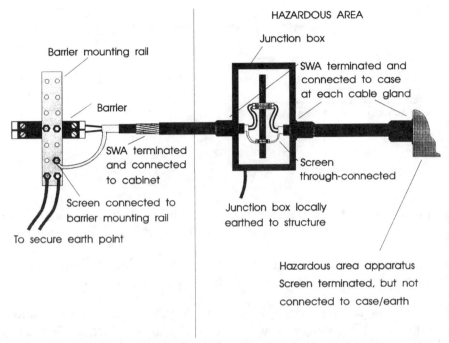

HAZARDOUS AREA

Junction box

Barrier mounting rail

SWA terminated and
connected to case
at each cable gland

Barrier

SWA terminated
and connected
to cabinet

Screen
through-connected

Screen connected to
barrier mounting rail

Junction box locally
earthed to structure

To secure earth point

Hazardous area apparatus
Screen terminated, but not
connected to case/earth

Figure 16.5   *General Earthing Arrangement for Intrinsically Safe Circuits*

busbar. This is because the barrier busbar earth conductor should
be segregated from cabinet earth until it reaches the main earth
point. That is, the earth connection from the barrier busbar should
be a dedicated insulated conductor of at least 4mm². A 6mm²
conductor is normal, and for installations containing large numbers
of barriers, 10mm² or 16mm² conductors are often required to
minimise volt drops which would otherwise affect measurement
sensitivity. The conductor used to connect between the barrier
busbar and the earth point should be such that the total earth loop
impedance between the barrier busbar, and the supply earth
(normally the neutral earth point) should be less than 1Ω.

Where the interconnecting cable contains screens to protect against
electrical interference, the screens should be earthed at one point
only. This is in any case a normal requirement if the screen is to

achieve its required function of protecting against interference. The earth point should be the earth busbar for the interface barrier. Figure 16.5 shows the overall position. As discussed in Chapter 15, it is useful if there are two separate insulated conductors used for this earth connection to facilitate periodic testing and maintenance. The connection points to both the barrier busbar and the system earth should be labelled 'SAFETY EARTH FOR INTRINSICALLY SAFE CIRCUITS: DO NOT REMOVE'.

Screens should, within the cable, be isolated one from another, and from any steel wire armour.

It should be noted that some countries have different approaches to earth bonding between the safe and hazardous area. If the plant installation is such that bonding conductors have already ensured there will be no potential difference between the different points of the plant, and that any earth fault currents which do flow will flow in the bonding system, then the importance of the 500V insulation for the electrical circuit in the hazardous area takes on less significance. Local recommendations and codes should always be consulted and followed.

Earthing of Electrical Apparatus Other than Intrinsic Safety

The CENELEC Standards EN50 014 et seq, clearly state a requirement for electrical apparatus with metallic enclosures to be provided with an external connection facility for an earthing or bonding conductor. This connection should be of increased safety design and be capable of accepting conductors of at least 4mm^2.[4] The CENELEC Standards also require an internal terminal, normally located close to other connection facilities for an earthing or bonding conductor. This facility should be able to accept conductors of similar size to the other connection conductors.[4][5] Apparatus certified to Standards other than CENELEC, for example to BS 4683, may or may not have a dedicated external earthing point. In either situation it is normal to ensure, as would be required by good engineering practice anyway, that any metallic case of such apparatus is electrically bonded to the local structure.

Screens and sheaths of cables may be treated in the same way as described for intrinsic safety.

General Earthing Arrangement for Hazardous Area Installation

Figure 16.6 shows the general earthing position for hazardous area installations, and identifies the differences between the safety earths and other earth connections.

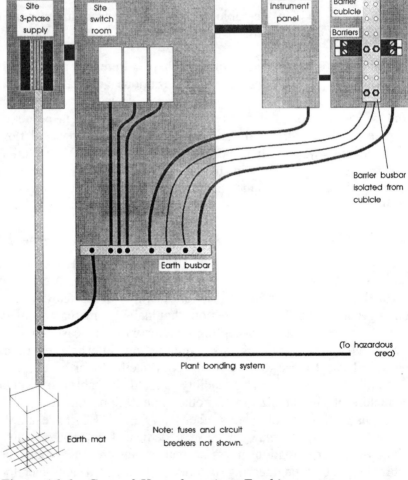

Figure 16.6 *General Hazardous Area Earthing*

Other points to note are:

Power systems should be earthed in accordance with normal regulations, for example IEE 'Regulations for Electrical Installations' in the UK.

If flexible conduit is used in an installation, the earth connections should not rely on the mechanical connection between different sections of conduit.

Electrical continuity between metallic enclosures, armour, SWA, conduit (except flexible conduit) should be achieved by the integrity of the mechanical joint itself. If additional external bonding is necessary, such bonding should be directly across the joint.

Potential Multiple Earth Arrangements

There is now an increasing use of potential multiple earthing practices, and on many installations, especially remote pipeline sites, PME earthing systems are used. The concept of PME is to deliberately earth the neutral conductor at each installation point, for example at each house in a street of houses. Although theoretical analysis of possible fault conditions indicates that there are numerous possibilities for dangerous situations to occur, in practice PME systems give no more trouble than other earth systems.

As far as hazardous area installations are concerned, there will normally be an isolation transformer at the power input to the site, and this should be suitably earthed to a local earth mat. This point, (or the connection to it) can then be regarded as the 'neutral earth point' and other connections to earth, for example from zener barrier installations, can be connected to this point.

The general principles are shown in Figure 16.7.

248

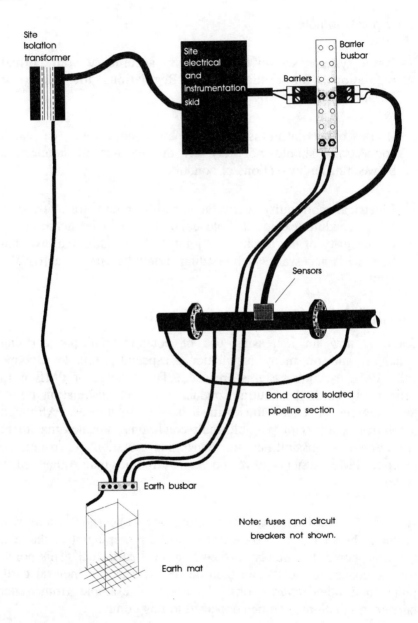

Figure 16.7 *Potential Multiple Earthing*

Pipeline monitoring remote stations normally require an isolated section of pipe, and the continuity of the pipe either side of the isolated section should be ensured by a bonding conductor which jumps the isolated section.

Notes and References

1 MTL's user Guide to Intrinsic Safety contains much useful advice on earthing of intrinsically safe circuits.

2 An article by G. Tortoishell entitled 'Earthing and Bonding - the potential for equalisation' published in Measurement and Control, Volume 19, September 1986 is useful further reading.

3 L.C. Towle's paper entitled 'Intrinsically Safe Installations on Ships and Offshore Structures' given to the Institute of Marine Engineers' meeting, London, October 1985 is a useful guide to the subject. It is available as Technical Paper number TP1074 from MTL.

4 This requirement is waived for situations such as double insulated electrical apparatus where earthing is not necessary, or in situations where earthing is not allowed.

5 Up to a maximum conductor size of 35mm^2.

CHAPTER 17

Inspection & Maintenance

Inspection & Maintenance

Regardless of how well protected electrical apparatus is when purchased, its continual security depends on good maintenance. Nothing stays in its purchased condition for long, and it is generally accepted that any complex piece of equipment, whether in private use, such as a car, or apparatus on an industrial plant, needs regular servicing to ensure that it will continue to perform the expected functions. As far as electrical apparatus in hazardous areas is concerned, there are two aspects which need consideration:

Will the apparatus perform the process function correctly?

Will the apparatus perform its process function within defined safety parameters relating to its presence in the hazardous area?

Very often the first aspect is well looked after. The user presumably needs the electrical apparatus to perform some task, and he will carry out such maintenance as is necessary to ensure this is achieved. However, very often the second aspect is overlooked, and much apparatus in hazardous areas which will work perfectly well (that is to say it will perform the process task intended), will not in fact be providing anything like the required level of safety from risk of ignition.

The situation is further complicated since, especially for such apparatus as luminaries and motors, the apparatus may be left well alone unless 'something goes wrong'. Whilst this lack of preventive maintenance may be acceptable from the process view point, it is not acceptable from the safety aspect, since if no maintenance is carried out until something goes wrong, it may be too late!

It is also very important to appreciate that approvals and certification for electrical apparatus relate to individual items of apparatus, and that these certifications can be totally invalidated by incorrect installation. So, in addition to the regular inspection and maintenance functions, there needs to be a very thorough check after installation, before the plant or apparatus is commissioned.

Once installation is satisfactorily complete, it will tend to be environmental conditions and unauthorised modifications which are the main causes of degradation.

This chapter considers the requirements for inspection and maintenance at both the installation (initial) and routine (periodic) stages.

It must be appreciated that the concern of this chapter is the safety from ignition caused by the electrical apparatus. The inspections and checks discussed are not for the purposes of operational functioning or general safety, and, although in some situations they may have a bearing on those aspects, additional inspection and testing tasks may be needed for operational and general purposes.

The Initial Inspection

Since it is vitally important to ensure that the equipment when installed still complies with its safety criteria, it is necessary to carry out an initial inspection in addition to any routine/periodic inspections or maintenance.

The purpose of the initial inspection is to ensure that the apparatus is indeed suitable for the hazardous area in which it has been installed, and that the act of installation has not in any way degraded the certification.

Among the key points which need to be established at this stage are:

A   Is the apparatus suitable for the hazardous area in which it has been installed? Especially consider:

    Temperature class   (marked on apparatus label)

Gas group   (marked on apparatus label)

Zone   (refer to method of protection)

Certification   (label should bear a certificate
                                 number)

Certification Authority   (marked on apparatus label)

Certification code   (marked on apparatus label)

B Is the apparatus suitable for the environmental conditions encountered at the point of installation? Especially consider:

Any environmental requirements if location needs superior protection to that automatically achieved by the method of protection. (IP codes or NEMA codes may be marked on the apparatus label, but it may be necessary to check with the manufacturer.)

C Will the installed apparatus maintain the required and intended environmental protection? Especially consider:

Cable entries (glands) to be correct to entry hole.

Cable used in gland to be suitable for gland.

Gland to be correctly assembled and tight.

Plant identification plates or other labels not fixed by methods which degrade environmental protection.

Fixing holes not drilled or positioned in such a way as to degrade environmental protection.

Any required gaskets of correct type and correctly positioned.

Any sealing compound of correct type and correctly applied. [1]

D Will the installed apparatus, maintain the essential requirements for the method of protection? Especially consider:

No modifications to supplied apparatus.

No unauthorised gaskets (especially on Ex d apparatus). [2]

No incorrect cable glands (especially on Ex d apparatus).

No incorrect cover or lid fixing bolts etc.

External earth stud (if fitted) correctly wired to local structure earth.

Any special electrical installation conditions (possibly indicated by letter 'X' at end of certificate number) have been complied with.

No hard setting compound used in flameproof joints.

No obstructions inside or outside flameproof joints.

Any required external current limiting or temperature limiting devices fitted and tested (especially Ex e).

Any pressure relief valves correctly operating (especially Ex p and Ex o).

Any filling medium present to required level (especially Ex o and Ex q).

No signs of leakage of filling medium (especially Ex o and Ex q).

Any associated safety interface or similar device present at correct location (especially Ex ia and Ex ib).

No infringement of creepage/clearance distances caused by stray wires of conductors at terminals (especially Ex N, Ex e, Ex ia/ib).

Any breathing and draining devices free from dirt, debris or obstructions.

Apparatus mounted in correct orientation (especially Ex o, Ex q).

Apparatus securely mounted.

No excessive pressure on associated cables and wires; especially at cable entry points, glands etc.

Any unused cable entry holes correctly blanked off.

Electrical connections tight.

Plant identification labels not degrading the integrity of the method of protection.

No risk of frictional sparks from reduced clearances on moving parts. (Especially consider fan blades on rotating machines.)

Correct fan and fan housing/guard fitted. (Especially check that plastic fan has not been fitted with metal cover.)

E   Is the cabling/wiring associated with the apparatus to a standard which will maintain the required environmental protection, protection against mechanical damage and protection against ignition? Consider especially:

Continuity of earth bonding.

Joints protected against rust and corrosion.

Conduit of correct type appropriate to the zone and method of protection with which it interconnects.

Stopper box/ sealing device used at any junction between hazardous area and non-hazardous area.

Stopper box etc. used if conduit enters flameproof/explosionproof apparatus.

Cables secured throughout their length.

No bends of unacceptably small radius in cables.

Cables not likely to be subject to excessive wear or damage caused by routing over sharp edges.

Adequate allowance made for vibration, for example flexible conduit or 'pig tail' end of cable used at entry to high vibration apparatus; pumps, motors, etc.

Any ducts, trenches etc. filled (e.g. with sand) or otherwise secured against passage of hazardous atmosphere.

Cable outer insulating sheath not damaged.

F   Plant marking present? Especially:

Apparatus tag number.

Cable idents.

G   Apparatus record made out and correct?

(See section later in this chapter.)

The above list is not exhaustive, and additional aspects, for example correct lamp ratings may need to be checked for some types of apparatus.

Clearly, these checks require the apparatus to be opened up, and thus the method of protection is void whilst the inspection is taking place. The apparatus and circuits must be isolated from the supply before any covers or lids are removed, or any action taken which will invalidate the protection concept. No live maintenance requiring access to the apparatus is permitted. [3] (There is an exception to this rule for intrinsically safe apparatus and circuits which will be considered later in the chapter.)

It is most important to appreciate that many of the checks listed above will be difficult to carry out once the apparatus is switched on and functional within the hazardous area. For some checks, to fully examine the apparatus, it may be necessary to remove some of the internal assembly, and this may be impossible or totally impractical after a functional stage is reached. For example, in order to examine if any fixing holes have been drilled through the base of a flameproof enclosure, the assembly within the enclosure may need to be taken out. Whilst it may not be necessary to go to such lengths in all situations, or for every item of apparatus, it would be prudent to check at least a sample of such enclosures on a new plant, where the labour used for the installation may not have been skilled or knowledgeable on the importance of the electrical installation from the hazardous area viewpoint. Indeed, the less the control or direct knowledge of the capabilities of the personnel carrying out installation work, the more important it is to carefully check every aspect which could affect safety. Thus, whatever temptation there may be, under the constraints and tight timescales of a plant start up to skimp or dispense with the initial inspection, it is absolutely essential that such an operation is indeed correctly carried out by competent personnel. Otherwise, the installation may be essentially unsafe and unprotected right from the start, and the faults leading to this condition may never be subsequently found.

The best time to carry out an initial inspection is immediately after installation, before the plant is operational and the location hazardous.

It will be appreciated that for the most part, the checks which are necessary take the form of a visual inspection, and in many cases all the necessary inspection can be carried out without the aid of any tools or special equipment apart from a screwdriver and whatever is needed to open the case or cover.

The requirement for an initial inspection is clearly stated in BS 5345: Part 1, with supplementary information in the detail parts of the Code for each method of protection.

In addition to carrying out the initial inspection at installation stage, the task should be repeated if apparatus is removed for repair etc. Clearly, some aspects may not be relevant on repeat initial inspections.

Electrical Tests at Installation (Initial Inspection) Stage

So far, the checks which have been considered are essentially visual inspections. There are also a number of simple electrical tests which need to be performed at this installation stage. Although the electrical tests are not complex, it is important that they are correctly carried out.

260

The electrical tests required are as follows:

Insulation resistance measurement

Most apparatus, especially that which is intrinsically safe, together with its associated wiring and cabling, should be capable of passing a 500 volt insulation resistance test to the frame or case of the apparatus. Check if the certificate has an 'X' condition, since this may refer to special requirements regarding insulation resistance or earth bonding of the case and circuit. [4]

It should be possible to achieve a reading well in excess of the requirement of 1 MΩ, and a value of 20 MΩ is normally perfectly realistic. In some cases, where the cable run is short, values of in excess of 200 MΩ may be appropriate. If the value obtained is lower than expected (even if it is greater than the required 1 MΩ) the apparatus and circuits should be examined to establish the reason since this may be a forewarning of impending problems on, for example, a poorly installed junction box or a cable with damp or damaged insulation. [5]

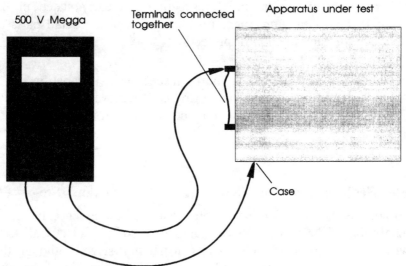

Figure 17.1   *500V Insulation Resistance Test on Apparatus*

The insulation resistance test should be carried out between the terminals of the apparatus, interconnected together, and the case of the apparatus. (Figure 17.1)

When testing interconnected cables, the disconnected cable should be tested core-to-core and cores-to-screen and to sheath/SWA etc.

In some situations, it may be possible to check the cabling and apparatus at one test from a suitable junction box or from the ends of the cable at the entry point to the non-hazardous area, without disconnecting the field apparatus.

Earth Tests

Normally, the case of the electrical apparatus in the hazardous area will be connected to local earth via its mounting. Apparatus which has been certified to the CENELEC Standards, having a metallic outer case, will normally have an external earth stud. This point should be connected to the local plant structure, using a suitable interconnecting lead. The termination to the plant structure should be arranged so that it will not vibrate loose.

If apparatus has a case made from plastic, or some other non conductive material, double insulation principles avoid the need for a local connection to earth.

Where screened cables have been employed, the screens should not normally be connected to earth (frame/case) in the hazardous area.[6]

Steel wire armoured cable and other cables with a conductive mechanical protection should normally have the armour connected to earth (usually via the cable glands) at each cable entry point.[6]

262

Many circuits, especially intrinsically safe circuits, need a safety earth connection at some point. The normal requirement in such situations is that the maximum earth loop impedance should not exceed 1Ω.[7] It is useful, at installation stage to measure both the earth loop impedance and the earth loop resistance. If these values are noted down, it will greatly assist any subsequent tests at the routine inspection stage.

Earth loop resistance tests can be easily achieved if two conductors are run to the earth point. (Figure 17.2)

Note that the two earth conductors are connected at separate (adjacent) points, to improve the integrity and ensure continued earth connection if one point corrodes or fails.

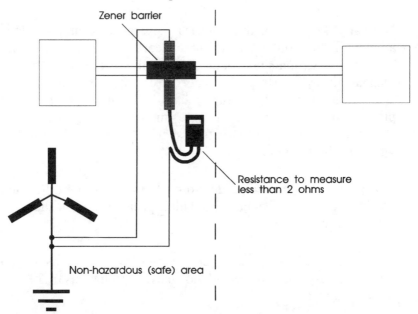

Figure 17.2   *Testing Earth Resistance Using Two Conductors*

With the two leads measured in series, a reading of less than 2Ω indicates that each conductor is less than 1Ω as required. (The two conductors in parallel will present less than 0.5Ω, but it should be possible to achieve less than 1Ω for each lead since this means that the earth connection will meet the requirements even if one lead is broken or a connection goes to high resistance.)

Other Tests

For purged apparatus, a test of the correct functioning of the surveillance circuit must be carried out prior to first use.

Periodic Inspections

Even if the apparatus is correctly installed initially, there is no guarantee that it will remain in this condition. The effects of environmental corrosion, vibration, unauthorised modifications and changes to plant and process layout and operation can all affect the safety and suitability of the electrical installation. It follows that in addition to the initial inspection, it is necessary to carry out routine or periodic inspections on all electrical apparatus and installations in the hazardous area.

The frequency of such inspections is a matter for determination by the responsible engineer on site. If the plant is mainly within a building, the process not subject to change, and the operating conditions clean and well ordered, routine inspections may only be necessary every 18 months or so. If, on the other hand, the plant is located outside in an exposed situation (especially if subject to corrosive elements such as salt spray or sand), or if process conditions give rise to apparatus becoming covered in deposits of resin, paint or chemicals, then the routine inspection will need to be very much more frequent: possibly every 3 months or so in some situations.

It is not intended that routine inspections should cause undue disturbance to the plant and processes. Whilst it may be necessary to isolate apparatus so that covers can be removed, it is often possible to carry out the majority of inspection tasks without opening apparatus, or even disconnecting the supply. Again, unless the circuits and apparatus are intrinsically safe, any operation which requires covers etc. to be removed must only be carried out after the circuit has been isolated, unless a gas free certificate or hot work permit has been issued.

Note.   The use of intrinsically safe test equipment such as intrinsically safe multimeters etc., does not mean that live testing can be carried out on apparatus which is not intrinsically safe. (Unless a hot work permit or gas free certificate has been issued.)

Amongst the key points which need to be checked at routine inspection are:

Apparatus tag number.

Cable idents correct to loop diagram/hook-up drawing.

Apparatus has no unauthorised modifications.

Any rectification work noted at previous inspections has been carried out suitably.

Earth connections secure.

No undue corrosion. (Especially on flanges for Ex d)

Cable entries tight.

No degradation of required IP rating.

No broken covers or fan cowls.

No build up of dirt on cooling fins. (Especially on motors)

Electrical connections tight. (Especially Ex e and Ex n)

Correct lamp ratings.

No changes to area classification. (If so, the method of protection or its gas group or T-class may not be suitable.)

No damage to associated cables.

No damage to apparatus.

Covers/lids correctly secured.

Apparatus mounting firm and acceptable.

Filters clean and free from dirt and debris.

Breathing and draining devices clean and free from dirt and debris.

Cable supports OK.

No external obstructions to flamepaths. (Ex d)

No excessive grease on flamepaths. (Ex d)

No hard setting compound on flamepaths. (Ex d)

No unauthorised gaskets. (Especially Ex d)

Gaskets of correct type.

No excessive grease on bearings.

No signs of excessive temperatures. (e.g. brittle or burnt insulation)

No cracks to ceramic feedthroughs or insulators. (Especially Ex d, Ex e, and Ex de)

No obvious changes to surrounding processes which could affect area classification.

No signs of leakage of filling medium. (Especially Ex o and Ex q)

Filling medium at required level. (Especially Ex o and Ex q)

No signs of leakage from stopper boxes or stopper glands. (Especially Ex d)

Electrical aspects remain secure: especially earth loop tests, and especially for Ex ia/ib. Compare value with previously noted value.

No dirt or obstructions to fan covers.

No signs of excessive vibration.

Surveillance circuits functioning correctly. (Especially Ex p)

Correct interface unit installed. (Ex ia/ib)

Conduit seals satisfactory at passage between nonhazardous area and hazardous area.

Cable idents correct, and no changes to wiring.

Again, the list above is not intended to be exhaustive, but should serve as a guide to the kind of check which is necessary. As for the initial inspection, the operation is mainly visual, but since the apparatus will need to be opened to check such aspects as tightness of connections and condition of flanges, it will be necessary to isolate the apparatus from the electrical supply unless a hot work permit or gas free certificate has been issued, or the apparatus concerned is intrinsically safe.

Use of Checklists

Although checklists can be very useful as a prompt for the type of check to do, they have the serious drawback that people tend to blindly follow them! It is not possible to anticipate all the faults which might occur, and even if one could do, the resulting list would be so long as to be of no practical use. There is always a temptation when following a check list to look at and check each item/task mentioned, and then to assume the work is complete. Any personnel who are carrying out inspections must be urged to stop and think once they reach the end of the check list to see if there are other aspects which need to be considered for any particular apparatus or location.

The more experienced the personnel at inspections, the less use will need to be made of checklists. However it is probably a good idea to have some basic plan put down on paper in the form of a general guideline and brief for the people involved.

Live Inspections/Maintenance with Intrinsic Safety

Unlike all other methods of protection, intrinsic safety allows for live maintenance and inspections. That is to say with intrinsic safety, the apparatus covers may be removed, and electrical connections checked, plugs and sockets opened etc. with the apparatus live and the area hazardous. Clearly, such activity may not always be desirable from an operational view point; for example, unless the appropriate precautions had been taken, it would not be advisable to start checking the tightness of terminals on a circuit relating to a primary shutdown system even if such activity was safe from the viewpoint of electrical ignition of the hazard.

268

The possibility of simple live maintenance tasks with intrinsic safety does however open the possibility of significant cost savings and reductions in nuisance value from down time of apparatus.

Records and Record Systems

There should be a record of the electrical apparatus located in the hazardous area, and any associated safe area apparatus, intrinsically safe interface units etc. The purpose of the record system is to enable management of an effective inspection system, for example reviewing the results of inspections to determine if the frequency of inspection can be extended or should be reduced.

Bearing in mind that very often, in spite of the certification requirement for certification labels to be legible and permanent, certification labels will be difficult to read after a period of years, it certainly makes good sense to ensure that all the certification details are recorded on the record card or system, together with details of the manufacturer, the plant tag number etc. An example of a record card is shown in Figure 17.3. The reverse of the card is used for information on inspections etc. Normally, only exceptions and faults observed during inspections are noted, together with details of any rectification work. Other than this, just the date and type of inspection (for example initial or routine) is noted. The value of earth loop impedance and resistance, and any insulation resistance test value should also be noted.

Although a hard copy record is perfectly acceptable, it is often more convenient to use a computer record, where details can be printed off in the form of a Job Sheet for any inspection work due. There are a number of suitable software packages available which will run on a personal computer, or networked into a larger system. The price of such software packages varies enormously, and very often the most costly are not the best value.

PLANT TAG NUMBER				
DESCRIPTION				
TYPE NUMBER		SERIAL NUMBER		
MANUFACTURER				
CERTIFICATION CODE		CERTIFICATE NUMBER		
CERTIFICATION AUTHORITY				
SPECIAL CONDITIONS				
HOOK UP/LOOP DRAWING				
CLASSIFICATION OF INSTALLED AREA				
GAS GROUP	ZONE	T-CLASS REQD		
ASSOCIATED APPARATUS/INTERFACES				
LOCATION/CABINET	DESCRIPTION		TAG NUMBER	
DATE INSTALLED		INSPECTION FREQUENCY		
* * * * * RECORD INSPECTIONS, REPAIRS, MODIFICATIONS, ETC. OVERLEAF * * * * *				

Figure 17.3 *Example of Record Form*

Surveys and Audits

Few plants stay unchanged for long, and it is worth bearing in mind that with changes to plant and process layout and function, the area classification may alter. Also the techniques used for the electrical installation may become out of date.

Regular reviews and audits of the plant are essential, even if the inspection procedures detailed in this chapter have been carried out. It is often helpful to use personnel from another division of the organisation, or an outside body to carry out such reviews so that a completely fresh view is obtained. There is always a tendency for personnel to miss aspects which are familiar to them, but which may in fact be the beginning of a dangerous situation, or signs of an unauthorised practice which, although happening for years has been overlooked by those responsible on the site.

It is difficult to recommend how often such surveys or audits should take place, since it will vary with the complexity of the plant, the general management and conditions of the site and the processes involved.

Portable Apparatus

As stated earlier, the use of portable intrinsically safe apparatus, test equipment etc. only permits live maintenance without a gas free certificate on intrinsically safe circuits. It should also be noted that unless specifically permitted by the certification documents, apparatus containing batteries should normally be removed to the non-hazardous area for replacement or charging.

Special Tools

It should be remembered that sparks can occur from tools being dropped or banged against stone or light alloy material. Personnel working in hazardous areas should clearly take care not to cause sparks by incorrect use of tools, and it may be wise, especially if working in elevated areas, gantries, towers etc., to tie tools to a personal safety line.

Phosphor bronze tools, or plastic coated tools (coated with a plastic which will not cause a static risk!) which will not cause spark ignition can be obtained for most purposes, but they tend to be expensive and are rarely used except in situations of particular risk, for example, hydrogen or acetylene atmospheres, and areas where work is necessary when flammable concentrations of hazard will definitely be present.

Notes and References

1 For information on the use of grease and sealing compounds with Ex d apparatus refer to Chapter 9.

2 For information on the use of gaskets and other sealing materials for flameproof protection 'd', refer to Chapter 9.

3 It is a standard requirement that a suitable means of isolating apparatus should be provided. This is discussed in more detail in Chapter 15.

4 For information on certification marking, refer to Appendix 1 and to the chapter dealing with a specific method of protection.

5 The 500 volt insulation test may be of special importance with intrinsically safe circuits, because such circuits may not use mechanically protected (e.g. SWA) cable. Thus the danger of incendive earth fault sparks is more likely than for other circuits where only mechanically protected cables are used.

6 For information on screens and sheaths of cable refer to Chapters 15 & 16.

7 For more information on earthing of intrinsically safe circuits refer to Chapter 16.

CHAPTER 18

Training

Training

Whilst it would appear that industrial training is gradually creeping towards the level of importance it deserves, it cannot be stressed too strongly that proper organised training of all personnel concerned with electrical apparatus in hazardous areas is essential for the continued safe operation of the plant.

Site visits and inspections of hazardous area electrical installations by the author invariably reveal glaring and disturbing deficiencies which seriously reduce the overall level of safety, and all too frequently the root cause of the problem is a lack of understanding by both management and operational personnel of the essential safety aspects for such installations.

Furthermore, the problems are not confined to small or less well established industries. Major operators too seem to experience real difficulty in ensuring that both their own personnel and outside contractors working on site have a sound working knowledge of this aspect of their work.

Electrical apparatus and its design, use and installation in hazardous areas, as well as the basic hazardous area layout are specialised topics. Although some advanced general education courses mention the subject in passing, few have either the time or the practical expertise to give the student even the basic information to lead to future competence. Thus sound industrial training leading to a competent level of understanding is essential for all personnel involved with hazardous areas.

It should be clear that before personnel carry out any kind of electrical work in hazardous areas, they should have a clear understanding of the methods of protection involved, and the requirements of safe installation, wiring etc. Even competent electrical fitters and technicians who are unfamiliar with hazardous areas, can easily leave installations in a state which falls far short of the safety levels needed.

Although the level and type of necessary training will vary according to the knowledge required and the work to be undertaken, the basic elements will need to be understood, together with any detail requirements relevant to the specific discipline, and at the very least, there should be a sound grasp of the fundamental principles of the subject by everyone involved in the subject.

Many organisations provide training of some form, but there are few places where a suitable total training package can be obtained, and some careful study of the requirements is well worthwhile before embarking on an exercise which is expensive both in time and money.

The following guidelines are not intended to provide a fixed syllabus, but should serve to outline the major subject headings which should be understood.

Those entries marked

* E * indicate that the training should include a written exercise

* P * indicate that training should include some practical work

* Q * indicate that a programmed or multiple choice question exercise should be included

MODULE A   General Training

It is important for everyone involved with hazardous areas, whether it be in supervisory, engineering, installation, or design work, to have a sound grasp of the different methods of protection and the certification procedures.

Module Content:

Introduction to hazardous areas, fires and explosions

The basics of area classification: terminology, gas groups, zones, temperature class, drawings

Responsibility of personnel, Health & Safety at Work Act, Codes of practice etc.

Summary of different methods of protection: 'd', 'e', 'n', 'p', 'm', 'q', 'o', 's', 'ib', 'ia'.

Certification authorities, types of certificate, labelling, component certification.

Main criteria for installation.

Portable apparatus, testing and inspection.

MODULE B Training for Designers

In addition to Module A, designers and specifiers will need additional training covering:

Revision and detail of Module A   * E * * Q *

Specification Standards and certification

Use of uncertified apparatus * Q *

Use of apparatus certified overseas * Q *

Selection and specification of apparatus depending on zone of installation * E *

Consideration of surface temperatures and ambient temperatures * Q *

Selection of compatible methods of protection * E *

System safety, especially 'ia' and 'ib' * E *

Possibilities for standardisation on items such as glands, junction boxes etc.

Environmental considerations * Q *

MODULE C   Inspection Personnel

In addition to Module A, those involved with site inspection work should complete the following:

Revision and detail of Module A   * E * * Q *

Cabling requirements * P *

Earthing requirements * E *

Terminations and connections * P *

Initial inspection of installations * E *

Earth testing and continuity * P *

Periodic inspections

Identification and labelling * E *

Mixtures of methods of protection * E *

Record of information

Typical faults and problems (see Chapter 17) * P *

Information from certificates

What to do if labels, certificates etc. are missing or unreadable

Use and abuse of check lists * E *

MODULE D   Installation Personnel

In addition to Module A, those involved with installation work should complete the following module. The depth of training will depend largely on previous experience and general level of installation knowledge.

Revision and detail of Module A   * E * * Q *

Cables:

> correct type * E *
> cable trays and ducts * E *
> protection of cables
> cleats and fixtures
> terminations
> sand traps and blocking of cable trays

Cable glands:

> correct type * E *
> environmental protection
> stress on gland
> correct tightening of glands * P *
> cable termination in glands * P *
> stopper glands * P *
> use of washers * E *
> electrical continuity of armour

Mounting of apparatus:

> manufacturers fixing points
> alternative fixings
> cable entries * Q *
> gaskets and sealing compounds * E *

Flameproof apparatus:

> additional holes in   * E *
> obstruction of flamepaths * E *

Rotating machines:

> vibration protection
> flexible conduit
> mixtures of methods of protection * E * or * Q *
> manufacturers instructions

Earthing:

> continuity of   * E *
> use of plugs and sockets * Q *
> checking and recording of results *E*
> cables for

Isolation devices:

> location and use * E *
> identification * Q *

Labelling:

> plant identification * Q *
> cable identification

MODULE E   Area Classification

In addition to Module A, personnel with responsibility for or involvement with area classification, should complete the following:

Revision and detail of Module A   * E * * Q *

Systems of area classification: IEC, North America etc.

Understanding of zone number and source of release

Understanding extents of zone: * E * * Q *

> LEL
> physical boundaries
> ventilation
> properties of hazard

Codes of practice: * E *

IEC 79-10
BS 5345: Parts 1 & 2
Institute of Petroleum Code
ICI/RoSPA Code
other specific industry codes, depending on industry concerned.

Examples of area classification

Use of area classification

Common sense classification * E *

Computations and equations * E *

Results and drawings * E *

Reviews * Q *

Exceptions and special considerations

Length of Training Courses

It has been tempting to include guideline durations for the training modules outlined above, but the depth of coverage required will, of course, vary considerably depending on the particular job function, site complexity, initial competence of the student etc. It may be worth noting, however, that excessively long courses may reflect the law of diminishing returns, and shorter, well organised courses with relevant subject matter may be preferable. This is especially important if the student has not been receiving regular training or study for a number of years, since such people will find it hard to concentrate effectively for more than one or two days. This observation tends to support the concept of short, pertinent courses, with frequent refresher sessions.

There is, of course, no substitute for experience in a well supervised situation, and this is as true for training as for work on the plant itself. If longer courses are contemplated, the amount of supervised practical work and detailed guidance given on the course should be carefully examined to ensure value for money.

Courses Run by Manufacturers

Many manufacturers of hazardous area electrical apparatus run excellent training courses. Such courses are, however, understandably geared towards their own products, and are unlikely to give a complete coverage of the subject. Most manufacturers who run such training courses regard the training as a silent sales activity, and even if the course is not in any way commercial in content, the passive intention is to gain commitment to a particular concept or product.

Thus, in general, training given by manufacturers should be regarded as supplementary to other training, and not as an easy alternative.

General Training Versus On Site Training

This is a difficult and complex discussion! If training is being given on site for a particular group of people from within one organisation, there is an enormous advantage that the training can be tuned to the precise needs of that site, and any problems or peculiar difficulties of that site can be fully explored. It must be remembered, however, that against this, some people are shy of appearing unknowledgable or foolish in front of their colleagues or managers, and thus may not ask questions about aspects which are bothering them or about which they are not clear. Thus any on site training should allow time in the schedule for each individual to have some one-to-one time with the trainer.

General or open courses have rather the reverse attributes. There is less time to divert on to specific problems which are unlikely to be of interest to the majority of students. However, people will generally be more willing to ask questions if they are among strangers rather than with people they work with and know well.

This observation is, however, perhaps less true of senior staff, where the opportunity to exchange ideas and discuss solutions to other people's problems tends to give added weight to outside training opportunities.

Very often, the decision of external general or on site course is dictated by cost, and if there are more than five or six people to be trained it is probably cost effective to hold the training on site.

Refresher Courses

Regular refresher courses should be held for all personnel. Those who are actively involved in hazardous area work most of the time can so often stray into poor practices, and those who encounter the subject infrequently will forget the important aspects amongst all the other subjects they need to command.

Training is an on-going discipline, and, as in all education, the best training will inspire the student to further study, and implant a degree of responsibility in the subject which will carry through to the day to day area of work.

As with virtually all technical subjects, the technology of electrical apparatus and hazardous areas is not static, and, as has been indicated throughout this book, the Standards and Codes are continually being reviewed. Practices which would have been considered perfectly satisfactory a few years ago may now be frowned upon, and again this highlights the need for constant training review.

There is little point in attempting to establish an exact frequency of training, but, as a final comment for thought, a day or two on a refresher course every one or two years seems a small price to pay for the benefits and additional safety which result.

APPENDICES

APPENDIX 1

Labelling

Labelling

The examples of labels in this appendix are intended to indicate the main information required by the different methods of protection. Actual details will vary from manufacturer to manufacturer and from one certifying authority to another.

The general form of CENELEC marking follows the convention:

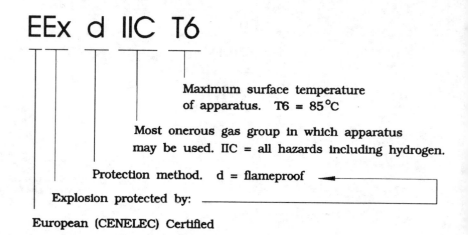

Figure A1.1 *CENELEC Marking Information*

Note on Certificate Numbers

Suffix letters at the end of a certificate number have the following meanings:

U   Component Certificate
X   Special conditions of use
B   Special conditions of use *
S   System Certificate *

*   No longer used, but found on some BASEEFA Certificates to national standards, eg Ex ia to SFA 3012.

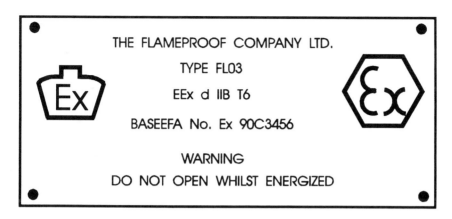

THE FLAMEPROOF COMPANY LTD.

TYPE FL03

EEx d IIB T6

BASEEFA No. Ex 90C3456

WARNING

DO NOT OPEN WHILST ENERGIZED

Figure A1.2 *Example of Label for Flameproof Apparatus Certified to European Standards*

Notes:

1 The apparatus has been certified by a certification organisation within the EEC which is listed in the EEC Directives. Hence the inclusion of the Distinctive Community Mark.

2 Note the warning label concerning live opening.

288

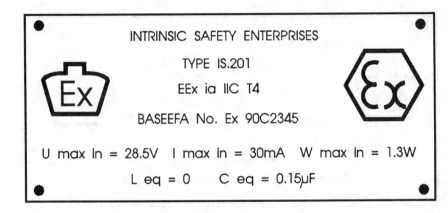

INTRINSIC SAFETY ENTERPRISES

TYPE IS.201

EEx ia IIC T4

BASEEFA No. Ex 90C2345

U max In = 28.5V   I max In = 30mA   W max In = 1.3W

L eq = 0   C eq = 0.15μF

Figure A1.3 *Example of Label for Intrinsically Safe Apparatus Certified to European Standards*

Notes:

1 The apparatus has been certified by a certification organisation within the EEC which is listed in the EEC Directives. Hence the inclusion of the Distinctive Community Mark.

2 The maximum input parameters U max in etc. are quoted. The interface unit used for this apparatus must ensure that these conditions are not exceeded.

3 The equivalent inductance and capacitance of the apparatus are quoted. This will affect cable parameters, and may affect combinations of several items of apparatus to be connected together.

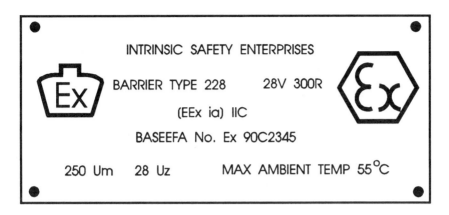

Figure A1.4 *Example of Label for an Intrinsic Safety Interface Unit Certified to the European Standards*

Notes:

1 The apparatus has been certified by a certification organisation within the EEC which is listed in the EEC Directives. Hence the inclusion of the Distinctive Community Mark.

2 The EEx ia is in brackets, indicating that it is Associated Apparatus. It is intended for use in a non-hazardous area.

3 Since it is to be located in the non-hazardous area, the surface temperature is not important from the view point of igniting the hazard. Thus there is no T class. Instead there is a maximum ambient temperature at which the safety interface can correctly perform its function. If no maximum ambient temperature is quoted, a 40°C ambient may be assumed.

4 The maximum safe input voltage Um is quoted. This is the maximum voltage which may be applied to the safe area terminals with the maximum output voltage of Uz still guaranteed.

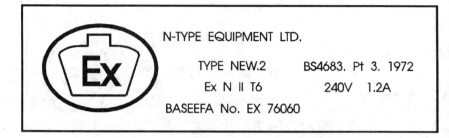

N-TYPE EQUIPMENT LTD.

TYPE NEW.2   BS4683. Pt 3. 1972

Ex N II T6   240V 1.2A

BASEEFA No. EX 76060

Figure A1.5   *Example of Apparatus Certified by BASEEFA to N-Type Protection of BS 4683.*

Notes:

1   The Standard number BS 4683: Part 3 is included.

2   N-type apparatus is normally acceptable for all gas sub groups, thus there is no sub group letter after the 'II'.

3   The apparatus is certified to a National Standard, so its code is Ex rather than EEx.

4   Note that the code letter is 'N'. BASEEFA have used both 'N' and 'n' in the past for apparatus Certified to BS 4683: Part 3, although the Standard calls for 'N'. The position is now clarified with the new Standard BS 6941, which will definitely use 'N'.

```
┌─────────────────────────────────────────────┐
│  THE COMPANY NAME                            │
│  MODEL GOOD.001                   (SA)®      │
│  1990                                        │
│  Max Ambient Temp 70°C                       │
│  Ex ia Class I, Groups A,B,C,D   T4A         │
│  INTRINSICALLY SAFE - SECURITE INTRINSEQUE   │
└─────────────────────────────────────────────┘
```

Figure A1.6 *Example of Label for Apparatus Approved by Canadian Standards Association*

Notes:

1   The labelling information uses the North American hazard classification system. However the 'Ex ia' follows the IEC system.

```
┌─────────────────────────────────────────────────┐
│  THE INTRINSIC SAFETY CO INC.  SEPULVEDOR CA      │
│                                                   │
│  ┌──────────┐                                     │
│  │ Factory  │        MODEL USA546                 │
│  │ Mutual   │        DRAWING ISC.1234.            │
│  │ System   │                                     │
│  └──────────┘                                     │
│  Approved                                         │
│  INTRINSICALLY SAFE FOR CLASS I DIVISION 1        │
│  GROUPS A,B,C,D.          T4A                     │
│  ENTITY PARAMETERS:                               │
│  V max 29v, I max 40mA,  CI 0.15µF,  LI 0         │
└─────────────────────────────────────────────────┘
```

Figure A1.7 *Example of Apparatus Approved by Factory Mutual Research Corporation (USA) using the Entity Approval Concept*

Notes:

1   The Entity Parameters indicate the maximum safe input levels, and hence assist the selection of a safe compatible interface unit. The Entity Parameters of this apparatus, in association with the Entity Parameters of the interface unit may be used to establish cable parameters.

2   The word 'Approved' after the Factory Mutual logo is optional.

```
┌─────────────────────────────────────────────┐
│  ╭───╮      The Explosionproof Company Inc.   │
│  │ U │                                         │
│  │® L│      Type XYZ   Serial No. 12345        │
│  ╰───╯                                         │
│        Class I Groups B C D  Divisions 1 & 2  │
│                                                │
│        Class II Groups E F                     │
│  Explosionproof    UL 674          NEMA 6      │
└─────────────────────────────────────────────┘
```

Figure A1.8   *Example of Label for Apparatus Approved (Listed) by
Underwriters Laboratories. USA.*

Notes:

1   The NEMA code refers to environmental protection. See
Appendix 7.

2   UL 674 refers to Electric Motors and Generators for use in
Hazardous (Classified) Locations.

APPENDIX 2

Minimum Ignition Curves

Minimum Ignition Curves

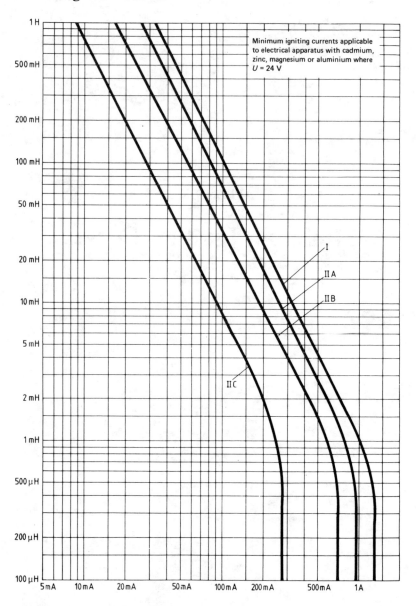

Minimum igniting currents applicable to electrical apparatus with cadmium, zinc, magnesium or aluminium where $U = 24\,V$

Figure A2.1 *Inductive Circuits*

297

Minimum Ignition Curves

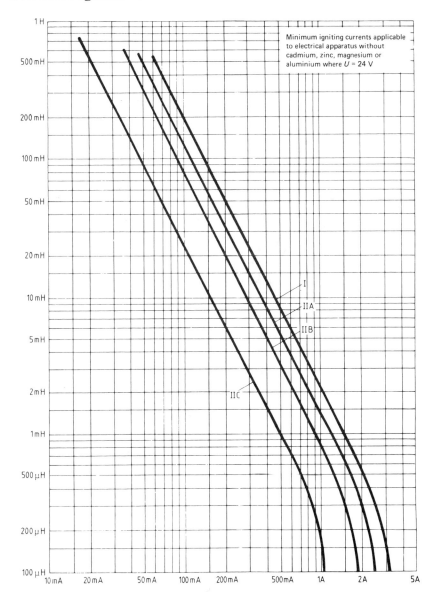

Figure A2.2 *Inductive Circuits*

298

Minimum Ignition Curves

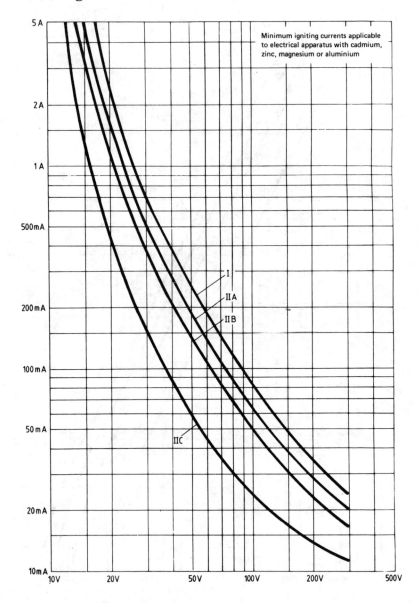

Figure A2.3 *Resistive Circuits*

Minimum Ignition Curves

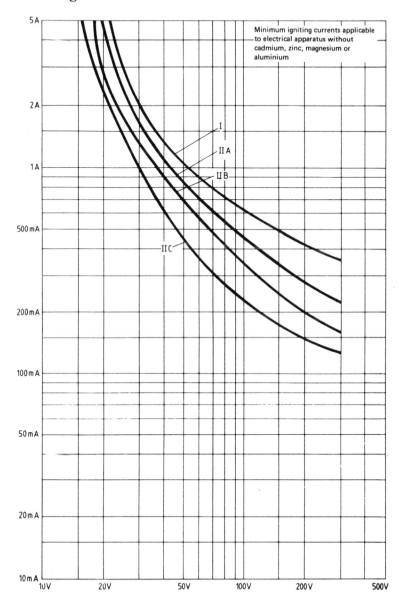

Figure A2.4 *Resistive Circuits*

300

Minimum Ignition Curves

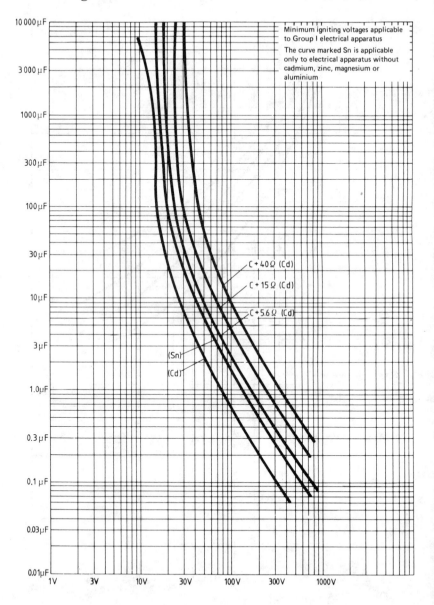

Figure A2.5 *Capacitive Circuits*

Minimum Ignition Curves

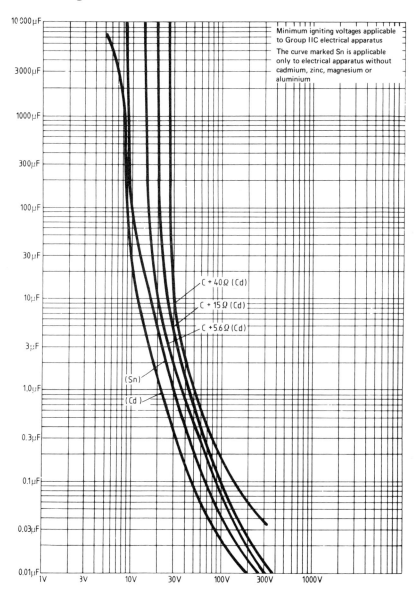

Figure A2.6 *Capacitive Circuits*

Creepage and Clearance Diagrams

304

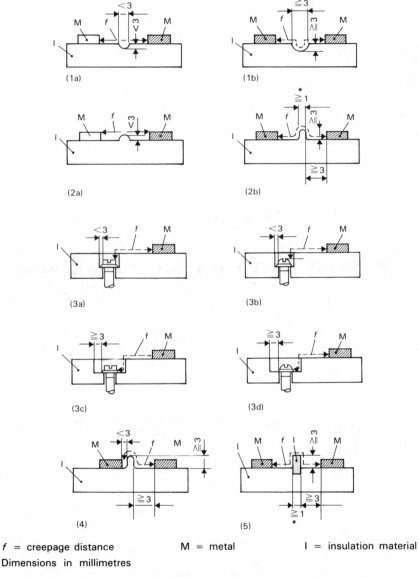

(1a)

(1b)

(2a)

(2b)

(3a)

(3b)

(3c)

(3d)

(4)

(5)

f = creepage distance   M = metal   I = insulation material
Dimensions in millimetres

Values marked * should be increased to 3mm where method of protection is increased safety unless certain other requirements are met.

Figure A3.1

Standards for Information and Installation

Standards for Information and Installation

There are a variety of Standards for different methods of protection which are intended primarily for the end user rather than the designer and manufacturer of apparatus.

The following list includes some of the more regularly used Standards, but is not intended to be exclusive. There are numerous Standards published in different countries and for particular industries. Some of these have been noted at appropriate points in the book.

Since the hazardous area classification system differs from country to country, especially between Europe and North America, the user should be careful to ensure that any Standard or Code of practice is relevant to the point of installation.

The IEC Standards form the root from which most other Codes of practice are based, and although in some respects they are more concerned with apparatus design, they are rarely used for full certification. Thus they are included in this section.

IEC STANDARDS

IEC 79- ELECTRICAL APPARATUS FOR EXPLOSIVE GAS ATMOSPHERES

IEC 79-0 PART 0 General requirements
Indicates general concepts, and defines gas groups, temperature classes etc.

IEC 79-1 PART 1 Construction and test of flameproof enclosures of electrical apparatus

IEC 79-1A First Supplement: Appendix D. Method of test for ascertainment of maximum experimental safe gap

IEC 79-2 PART 2 Electrical apparatus-type of protection 'p'

307

IEC 79-3	PART 3	Spark test apparatus for intrinsically safe circuits
IEC 79-4	PART 4	Method of test for ignition temperature
IEC 79-5	PART 5	Sand-filled apparatus
IEC 79-6	PART 6	Oil-immersed apparatus
IEC 79-7	PART 7	Construction and test of electrical apparatus, type of protection 'e'
IEC 79-10	PART 10	Classification of hazardous areas
IEC 79-11	PART 11	Construction and test of intrinsically safe and associated apparatus
IEC 79-12	PART 12	Classification of mixtures of gases or vapours with air according to their maximum experimental safe gaps and minimum igniting currents
IEC 79-13	PART 13	Construction and use of rooms or buildings protected by pressurization
IEC 79-14	PART 14	Electrical installations in explosive gas atmospheres (other than mines) A general Code of practice. Defines the methods of protection suitable for different zones. States the main requirements for the installation of apparatus.
IEC 79-15	PART 15	Electrical apparatus for explosive gas atmospheres, type of protection 'n'.

BRITISH STANDARDS

BS 5345   Code of practice for selection, installation and maintenance of electrical apparatus for use in potentially explosive atmospheres (other than mining applications or explosives processing and manufacturer)

BS 5345   PART 1   General recommendations

BS 5345   PART 2   Classification of hazardous areas

BS 5345   PART 3   Installation and maintenance requirements for electrical apparatus with type of protection 'd'. Flameproof enclosure

BS 5345   PART 4   Installation and maintenance requirements for electrical apparatus with type of protection 'i'. Intrinsically safe electrical apparatus and systems.

BS 5345   PART 5   Installation and maintenance requirements for electrical apparatus protected by pressurization 'p' and by continuous dilution, and for pressurized rooms.

BS 5345   PART 6   Installation and maintenance requirements for electrical apparatus with type of protection 'e'. Increased safety.

BS 5345   PART 7   Installation and maintenance requirements for electrical apparatus with type of protection N.

BS 5345   PART 8   Installation and maintenance requirements for apparatus with type of protection 's'. Special protection.

BS 5345   PART 9 * Installation and maintenance requirements for electrical apparatus with type of protection 'o'. Oil-immersed apparatus, and with type of protection 'q'. Sand-filled apparatus

* Not yet published.

NOTE: Copies of British and IEC Standards may be obtained from:

The British Standards Institution
Linford Wood
Milton Keynes
MK14 6LE

Telephone 0908 220022
Telex 825777
Fax 0908 320856

USA

INSTRUMENT SOCIETY OF AMERICA RECOMMENDED PRACTICE CODES

RP12.1   Electrical Instruments in Hazardous Atmospheres

RP12.2   Intrinsic Safety

RP12.3   Explosion-Proof

RP12.4   Instrument Purging for Reduction of Hazardous Area Classification

RP12.5   Sealing and Immersion

RP12.6   Installation of Intrinsically Safe Systems for Hazardous (Classified) Locations

S12.10   Area Classification in Hazardous (Classified) Dust Locations
Adopted by ANSI 1988

S12.11   Electrical Instruments in Hazardous Dust Locations

S12.12   Electrical Equipment for use in Class I, Division 2 Hazardous (Classified) Locations
Adopted by ANSI 1984

310

AMERICAN NATIONAL STANDARDS INSTITUTE /
NATIONAL FIRE PROTECTION ASSOCIATION

ANSI/NFPA 70   NATIONAL ELECTRICAL CODE

Canada

National Electrical Code.

Standards for Design and Certification

312

Standards for Design and Certification

There are numerous Standards which are used for the design and certification of electrical equipment. The historical development of the methods of protection and the harmonising of Standards, especially within Europe, has left a complex picture, and it will be many years before apparatus to the older Standards ceases to be available. It should be emphasised that Ex certification (as opposed to EEx) is perfectly acceptable from the safety in hazardous areas aspect, but the implication of certification to a National Standard may be important to countries other than that in which the apparatus was certified.

There is no doubt that within Europe, and especially within the EEC, the trend is towards the use of CENELEC certification Standards, although at the time of writing some National Standards are still used.

The following list is not fully comprehensive, but should serve as a general guide to the more common Standards which may be encountered.

PROTECTION METHOD	CODE	CENELEC	BRITAIN	GERMANY	FRANCE	USA	CANADA	AUSTRALIA
GENERAL INFORMATION		EN 50 014	BS 5501 Part 1	VDE0170/0171-1	NF C23-514	NFPA-70	NEC	AS 2380 Part 1
FLAMEPROOF (EXPLOSIONPROOF)	EEx d Ex d Ex d Ex d	EN 50 018	BS 5501 Part 5 BS 4683 Part 2 BS 229 BS 5000 Part 17**	VDE0170/0171-5	NF C23-518	UL1203 UL674	-	- AS 2480
INCREASED SAFETY	EEx e Ex e Ex e	EN 50 019	BS 5501 Part 6 BS 4683 Part 4 BS 5000 Part 15**	VDE0170/0171-6	NF C23-519	-	-	-
N-TYPE*	Ex n or N Ex N Ex N		BS 4683 Part 3 BS 6941 BS 5000 Part 16**					AS 2238
INTRINSIC SAFETY ***	EEx ia/ib Ex ia/ib Ex ia/ib	EN 50 020	BS 5501 Part 7 SFA 3012 BS 1259	VDE0170/0171-7	NF C23-520	- FM3610 UL913	- CSA22.2-157	- AS 2380 Part 7
PRESSURIZED	EEx p Ex p	EN 50 016	BS 5501 Part 3	VDE0170/0171-3	NF C23-516	NFPA 496	-	AS 1825
ENCAPSULATION	EEx m Ex m	EN 50 028	BS 5501 Part 8	VDE0170/0171-8	NF C23-528	-	-	- AS 2431
SPECIAL PROTECTION	Ex s		SFA 3009					AS 1826
OIL FILLING	EEx o Ex o	EN 50 015	BS 5501 Part 2	VDE0170/0171-2	NF C23-515	-	-	-
SAND/POWDER FILLING	EEx q	EN 50 017	BS 5501 Part 4	VDE0170/0171-4	NF C23-517	-	-	-

* BS 6941 refers to type of protection N, whereas IEC 79-15 refers to type of protection 'n'. The method of protection (non sparking) is, with minor exceptions, the same in both case.

** Standard for rotating electrical machines. Note that compliance with this Standard does not necessarily mean that it is certified.

*** See also EN 50 039 (BS 5501: Part 9) for intrinsically safe systems.

APPENDIX 6

Certification and Standards Bodies

Certification and Standards Bodies

SECTION 1: EEC APPROVED TEST BODIES

At the time of going to press, the organisations marked with an asterisk (*) are cited in the EEC Directives, and apparatus certified by them may bear the Distinctive Community Mark

in addition to their own Approved Test Body mark.

NOTE

In the information which follows, telephone numbers and fax numbers are given complete with dialing codes from the UK. From outside the UK, delete the part in parentheses ( ) and replace with the appropriate international code.

AUSTRIA

ETI Bundesversuchs-und Forschungsantalt Arsenal
Electrotechnisches Institut
ETI
Abt. Elektnsche Sicherhiet
Faradaygasse 3
Postfach 8
A-1030 Wein
Austria
Telephone (010 43) 1 7825310

TUV Technischer Uberwachungsverein
TUV
Krugerstrasse 16
A-1010 Wein
Austria
Telephone (010 43) 1 51407

BELGIUM

INIEX   Institut National des Industries Extractives
    * INIEX
    60 Rue Grande
    B7260 Colfontaine
    Belgium
    Telephone   (010 32) 65 67 2343
    Fax   (010 32) 65 66 0953

DENMARK

DEMKO   Danmarks Elektriske Materialkontrol
    * DEMKO
    Lyskaer 8
    2730 Herlev
    Denmark
    Telephone   (010 45) 42 94 7266
    Fax   (010 45) 42 94 7261

FRANCE

CERCHAR Centre d'Etudes et Recherches des Charbonnage de
    France
    * CERCHAR
    Laboratoire de Verneuil en Halatte
    Boite Postale No 27
    60103 Creil
    France
    Telephone   (010 33) 44 55 6677
    Fax   (010 33) 44 55 6699

LCIE   Laboratoire Central des Industrial Electriques
* LCIE
33 Avenue du General Leclerc
92260 Fontenay-aux-Roses
France
Telephone (010 33) 40 95 6060
Fax   (010 33) 40 95 6095

GERMANY

PTB   Physikalisch-Technische Bundesanstalt
* PTB
Bundesalle 100
D-3300 Braunschweig
Germany
Telephone (010 49) 531 5920
Fax   (010 49) 531 5924006

BVS

    * BVS
Beylingstrasse 65
D-4600 Dortmund 14
Germany
Telephone (010 49) 231 2491226
Fax   (010 49) 231 2491224

ITALY

CESI   Centro Elettrotecnico Sperimentale Italiano
* CESI
Via Rubattino 54
I-20134 Milano
Italy
Telephone (010 39) 2 21251
Fax   (010 39) 2 2125440

NORWAY

NEMKO   Norges Elektnske Materiellkontrol
NEMKO
Gaustadallen 30
Postboks 288
Blindern
Oslo 3
Norway
Telephone   (010 47) 2 691950
Fax   (010 47) 2 698636

SPAIN

LOM   Laboratorio Oficial Jose Maria Madariaga
* LOM
C/Alenza
1y2-Madrid-3
Spain
Telephone   (010 34) 1 442 1987
Fax   (010 34) 1 441 9933

SWEDEN

SP   Statens Provningsanstalt
SP
Brinellgaten 12
Box 857
51015 Boras
Sweden
Telephone   (010 46) 33 165000
Fax   (010 46) 33 135502

320

SWITZERLAND

SEV   Schweizenscher Elektrotechnischer Verein
             SEV
             Seefeldstrasse 301
             CH-8008
             Zurich
             Switzerland
             Telephone   (010 41) 1 384 9111
             Fax   (010 41) 1 551426

UK

             Electrical Equipment Certification Service

BASEEFA   British Approvals Service for Electrical Apparatus in
          Flammable Atmospheres
             * BASEEFA
             Health & Safety Executive
             Harpur Hill
             Buxton
             Derbyshire
             SK17 9JN
             Telephone   0298 26211
             Fax   0298 79514

MECS   Mining Equipment Certification Service
             * MECS
             Harpur Hill
             Buxton
             Derbyshire
             SK17 9JN
             Telephone   0298 26211
             Fax   0298 79514

SCS   Sira Certification Services Limited

* SCS
Saighton Lane
Saighton
Chester
CH3 6EG

Telephone 0244 332200
Fax   0244 332112

SECTION 2: OTHER TEST AND APPROVALS ORGANISATIONS

CZECHOSLAVAKIA

Prufstelle Nr 214
co VVUU
71607 Ostrava-Radvanice

Telephone (010 42) 69 215444
Fax   (010 42) 69 222502

YUGOSLAVIA

Komisija Za Ispitivanje S-Uredaja

S-Commission
41001 Zagreb
Postanski Pretinac 304
Yugoslavia

Telephone (010 38) 41 312222

322

CANADA

CSA   Canadian Standards Association

CSA
178 Rexdale Boulevard
Rexdale
(Toronto)
Ontario
Canada
M9W 1R3

Telephone (0101) 416 744 4145
Fax   (0101) 416 747 4149

USA

UL   Underwriters Laboratories

UL
333 Pfingsten Road
Northbrook
Ilinois 60062
USA

Telephone (0101) 312 272 8800
Fax   (0101) 312 272 8129

FM   Factory Mutual Research

FM
1151 Boston Providence Turnpike
Norwood
MA 02062
USA

Telephone (0101) 617 762 4300
Fax   (0101) 617 762 9375

323

AUSTRALIA

SA Standards Australia

SA
Standards House
80 Arthur Street
North Sydney
NSW
Australia

Telephone (010 61) 2 963 4111
Fax   (010 61) 2 959 3896

SOUTH AFRICA

SABS South African Bureau of Standards

SABS
Private Bag X 191
Pretoria 0001
Republic of South Africa

Telephone (010 27) 12 48 1311

SECTION 3: OTHER STANDARDS ORGANISATIONS

BSI British Standards Institution
2 Park Street
London
W1A 2BS

ISA Instrument Society of America
67 Alexander Drive
PO Box 12277
Research Triangle Park
NC 27709
USA

ANSI   American National Standards Institute
1430 Broadway
New York 10018
USA

NFPA   National Fire Protection Association
Batterymarch Park
Quincy
MA 02269
USA

Environmental Protection

There are two systems in common use for defining the amount of protection an enclosure affords against the ingress of dust and liquids. The main system is the IP Code. This is defined fully in IEC 529. The Standard is reproduced by BSI as BS 5490, 1985, and the information is also given in the Code of practice, BS 5345: Part 1. The NEMA Code system is often used in North America as an alternative to the IP Code.

The IP code uses two digits to specify environmental protection. The first digit signifies the protection against solid matter (dust etc.) and the second digit specifies the protection against liquid (water).

FIRST NUMERAL	DEGREE OF PROTECTION
0	No protection of persons against contact with live or moving parts inside the enclosure. No protection of equipment against ingress of solid foreign bodies.
1	Protection against accidental or inadvertent contact with live or moving parts inside the enclosure by a large surface of the human body, for example, a hand, but no protection against deliberate access to such parts. Protection against ingress of large solid foreign bodies.
2	Protection against contact with live or moving parts inside the enclosure by fingers. Protection against ingress of medium sized solid foreign bodies
3	Protection against contact with live or moving parts inside the enclosure by tools, wires or such objects of thickness greater than 2.5mm. Protection against ingress of small solid foreign bodies.
4	Protection against contact with live or moving parts inside the enclosure by tools, wires or such objects of thickness greater than 1mm. Protection against ingress of small solid foreign bodies.
5	Complete protection against contact with live or moving parts inside the enclosure. Protection against harmful deposits of dust. The ingress of dust is not totally prevented, but dust cannot enter in an amount sufficient to interfere with satisfactory operation of the equipment enclosed.
6	Complete protection against contact with live or moving parts inside the enclosure. Protection against ingress of dust.
SECOND NUMERAL	DEGREE OF PROTECTION
0	No protection.
1	Protection against drops of water. Drops of condensed water falling on the enclosure shall have no harmful effect.
2	Protection against drops of liquid. Drops of falling liquid shall have no harmful effect when the enclosure is tilted at any angle up to 15° from its normal position.
3	Protection against rain or sprayed water. Water falling in rain at an angle of not more than 60° from the vertical shall have no harmful effect.
4	Protection against splashing. Liquid splashed from any direction shall have no harmful effect.
5	Protection against water jets. Water projected by a nozzle from any direction shall have no harmful effect.
6	Protection against conditions on ship's decks. (Deck watertight equipment) Water from heavy seas shall not enter the enclosure under prescribed conditions.
7	Protection against immersion in water: it shall not be possible for water to enter the enclosure under stated conditions of pressure and time.
8	Protection against indefinite immersion in water under specified pressure. It shall not be possible for water to enter the enclosure.

An alternative to the IP code is the NEMA code, which is used extensively in North America. The NEMA code is used to specify both environmental protection and hazardous area suitability. Full details in ICS 1-110.

ENCLOSURES FOR INDOOR NON-HAZARDOUS LOCATIONS

PROVIDES PROTECTION AGAINST	TYPE OF ENCLOSURE							
	1	2	4	4X	6	11	12	13
Accidental contact with enclosed equipment	yes	yes	yes	yes	yes	yes	yes	yes
Falling dirt	yes	yes	yes	yes	yes	yes	yes	yes
Falling liquids and light splashing	-	yes	yes	yes	yes	yes	yes	yes
Dust, lint, fibres and flyings	-	-	yes	yes	yes	-	yes	yes
Hosedown and splashing water	-	-	yes	yes	yes	-	-	-
Oil and coolant seepage	-	-	-	-	-	-	yes	yes
Oil and coolant spraying and splashing	-	-	-	-	-	-	-	yes
Corrosive agents	-	-	-	yes	-	yes	-	-
Occasional submersion	-	-	-	-	yes	-	-	-

ENCLOSURES FOR OUTDOOR NON-HAZARDOUS LOCATIONS

PROVIDES PROTECTION AGAINST	TYPE OF ENCLOSURE			
	3	3R	3S	6
Accidental contact with enclosed equipment	yes	yes	yes	yes
Rain snow and sleet	yes	yes	yes	yes
Sleet	-	-	yes	-
Windblown dust	yes	-	yes	yes
Hosedown	-	-	-	yes
Occasional submersion	-	-	-	yes

ENCLOSURES FOR INDOOR HAZARDOUS LOCATIONS

PROVIDES PROTECTION AGAINST ATMOSPHERES CONTAINING	TYPE OF ENCLOSURE									
	CLASS	GROUP	7A/8A	7B/8B	7C/8C	7D/8D	9E	9F	9G	10
Acetylene	I	A	yes	-	-	-	-	-	-	-
Hydrogen, manufactured gas	I	B	yes	yes	-	-	-	-	-	-
Ethyl ether, ethylene, cyclopropane	I	C	yes	yes	yes	-	-	-	-	-
Gasoline, hexane, naphtha, benzine, butane, propane, alcohol, acetone, natural gas, laquer solvent, benzol	I	D	yes	yes	yes	yes	-	-	-	-
Metal dust	II	E	-	-	-	-	yes	-	-	-
Carbon black, coal dust, coke dust	II	F	-	-	-	-	yes	yes	-	-
Flour, starch, grain dust	II	G	-	-	-	-	yes	yes	yes	-
Fibres, flyings	III		-	-	-	-	yes	yes	yes	-
Methane, with or without coal dust	Mines		-	-	-	-	-	-	-	yes

NOTES

Type 7 and Type 10 are 'Explosion Proof' enclosures.
Type 8 is Oil Filled enclosures.
Type 9 is Dust Explosion Proof.
If the installation is outdoors and/or additional protection against environmental conditions is required, a combination-type enclosure will be needed. Enclosures which meet the requirements of more than one type of enclosure may be designated by a combination of type numbers, the smaller number being given first.

Properties of Some Flammable Materials[1] [2]

FLAMMABLE MATERIAL	MELTING POINT °C	BOILING POINT °C	RELATIVE VAPOUR DENSITY	FLASH POINT °C	FLAMMABLE LIMITS % VOL UEL	FLAMMABLE LIMITS % VOL LEL	AUTO[3] IGNITION TEMP °C	T-CLASS[3]	APPARATUS GAS GROUP	MOLECULAR (FORMULA) WEIGHT	VAPOUR PRESSURE (Pa) 25°C	VAPOUR PRESSURE (Pa) 40°C
acetaldehyde	-123	20	1.52	-38	57	4	140	T4	IIA	44.05		
acetic acid	17	118	2.07	40	16	5.4	485	T1	IIA	60.05		
acetic anhydride	-73	140	3.52	54	10	2.7	(334)	(T2)	IIA	102.09		
acetone	-95	56	2.0	-19	13	2.15	535	T1	IIA	58.08	30785	56593
acetonitrile	-45	82	1.42	5	4.4		523	T1	IIA	41.05		
acetyl chloride	-112	51	2.7	4			390	T2	IIA	78.5		
acetylene	-81	-84	0.9		100	5.0	305	T2	IIC	26.02		
acrylonitrile (4)	-82	77	1.83	-5	17	1.5	480	T1	IIA	53.06		
allyl alcohol				21		3			IIA	58.08		
allyl chloride	-135	45	2.64	-20	11.2	3.2	485	T1	IIA			
allylene	-103	-23	1.38			1.7			IIB			
ammonia	-78	-33	0.59		28	15	630	T1	IIA			
amphetamine		200	4.67	<100					IIA			
aniline	-6	184	3.22	75	8.3	1.2	617	T1	(IIA)	93.13		
benzaldehyde	-26	179	3.66	65		1.4	190	T4	(IIA)	106.12		
benzene	6	80	2.7	-11	8	1.2	560	T1	IIA	78.11	12689	24369
blast furnace gas					70	28			IIA			
1-bromobutane	-112	102	4.72	<21		2.5	265	T3	IIA	137.03		
bromoethane	-119	38	3.75	<-20	11.3	6.7	510	T1	IIA	108.97		
buta-1, 3-diene	-109	-4	1.87		12.5	2.1	430	T2	IIA	54.09		
butane	-138	-1	2.05	-60	8.5	1.5	365	T2	IIA	58.12		
butanone (MEK)	-86	80	2.48	-1	11.5	1.8	505	T1	IIA	72.11		
butan-1-ol	-89	118	2.55	29	9	1.7	340	T2	IIA	74.12		2359
butyl acetate	-77	127	4.01	22	8	1.4	370	T2	IIA	116.16		
butylamine	-104	63	2.52	-9			(312)	(T2)	IIA	73.14		
but-1-ene	-185	-6	1.95		10	1.6	385	T2	IIA			
carbon disulphide	-112	46	2.64	-20	60	1	102	T5	IIC			
carbon monoxide	-205	-191	0.97		74.2	12.5	605	T1	IIA			
chlorobenzene	45	132	3.88	28	7.1	1.3	637	T1	IIA	112.56	13612	25977
1-chlorobutane	-123	78	3.2	<0		1.8	(460)	(T1)	IIA	92.57		
chloroethane	-136	12	2.22		15.4	3.6	510	T1	IIA	64.52		
2-chloroethanol	-70	129	2.78	55	16	5	425	T2	IIA	80.52		
chloroethylene	-154	-14	2.15		29.3	3.8	470	T1	IIA	62.5		
chloromethane	-98	-24	1.78		13.4	10.7	625	T1	IIA	50.49		
chloromethyl methyl ether	-103	60							IIA			

FLAMMABLE MATERIAL	MELTING POINT °C	BOILING POINT °C	RELATIVE VAPOUR DENSITY	FLASH POINT °C	FLAMMABLE LIMITS % VOL		AUTO IGNITION TEMP °C	T-CLASS	APPARATUS GAS GROUP	MOLECULAR (FORMULA) WEIGHT	VAPOUR PRESSURE (Pa)	
					LEL	UEL					25°C	40°C
1-chloropropane	-123	37	2.7	-18	2.8	10.7	(592)	(T1)	IIA	78.54		
2-chloropropane		47	2.7	-32	2.6	11.1	520	T1	IIA	78.54		
cresol	11	191	3.73	81	1.1		555	T1	IIA	108.14		
crotonaldehyde	-75	102	2.41	13	2.1	15.5	(230)	(T3)	IIB	70.09		
cumene	-97	152	4.13	36	0.88	6.5	420	T2	IIA			
cyclobutane	-91	13	1.93		1.8				IIA	56.1		
cycloheptane		119	3.39	<21					IIA	98.18		
cyclohexane	7	81	2.9	-18	1.2	7.8	259	T3	IIA	84.16	13018	24637
cyclohexanol	24	161	3.45	68	1.2		300	T2	IIA	100.16		
cyclohexanone	-31	156	3.38	43	1.4	9.4	419	T2	IIA	98.15		
cyclohexene	-104	83	2.83	<-20	1.2		(310)	(T2)	IIA	82.15		
cyclohexylamine	-18	134	3.42	32			290	T3	IIA	99.18		
cyclopentane	-93	47		-37			(380)	(T2)	IIA	70.13		
cyclopropane	-127	-33	1.45		2.4	10.4	495	T1	IIA	42.08		
decahydronaphthalene	-43	196	4.76	54	0.7	4.9	260	T3	IIA	138.26		
decane	-30	173	4.9	96	0.8	5.4	205	T3	IIA	142.29		
dibutyl ether	-95	141	4.48	25	1.5	7.6	185	T4	IIB	130.22		
dichlorobenzene	-18	179	5.07	66	2.2	9.2	(640)	(T1)	IIA	147.01		
1,1-dichloroethane	-98	57	3.42	-10	5.6	16	440	T2	IIA	98.96	10518	20631
1,2-dichloroethane	-36	84	3.42	(5)	6.2	15.9	(413)	(T2)	IIA	98.96	80185	135206
1,1-dichloroethylene	-122	37	3.4	-18	7.3	16	(570)	(T1)	IIA	96.94		
1,2-dichloroethylene	<-80	33	3.55	-10	9.7	12.8	(440)	(T2)	IIA	96.94		
1,2-dichloropropane		96	3.9	15	3.4	14.5	555	T1	IIA	112.99		
diethyl ether	-116	34	2.55	<-20	1.7	36	170	T4	IIB	74.12		
diethyl sulphate	-25	208	5.31	104					IIA	154.18		
diethylamine	-50	56	2.53	<-20	1.7	10.1	(310)	(T2)	IIA	73.14		
dihexyl ether	-43	227	6.43	75			185	T4	IIA			
di-isobutylene	-106	105	3.87	(2)			(305)	(T2)	IIA	112.22		
di-isopropyl ether	-86	69	3.52	-28	1.4	21	(416)	(T2)	IIB	102.17		
dimethyl ether	-141	-25	1.59		3.7	27	(400)	(T2)	IIB	46.07		
dimethylamine	-92	7	1.55		2.8	14.4	(440)	T2	IIA			
dimethylformamide	-61	152	2.51	58	2.2	15.2	379		IIA	73.10		1339
1,4-dioxane	10	101	3.03	11	1.9	22.5			IIB	88.10		
1,3-dioxolane	-26	74	2.55	(2)					IIB	74.08	4979	10252
dipentyl ether	-69	170	5.45	(57)			170	T4	IIA	158.29		
dipropyl ether	-122	90	3.53	<21			170	T4	IIB			

332

FLAMMABLE MATERIAL	MELTING POINT °C	BOILING POINT °C	RELATIVE VAPOUR DENSITY	FLASH POINT °C	FLAMMABLE LIMITS % VOL LEL	FLAMMABLE LIMITS % VOL UEL	AUTO IGNITION TEMP °C	T-CLASS	APPARATUS GAS GROUP	MOLECULAR (FORMULA) WEIGHT	VAPOUR PRESSURE (Pa) 25°C	VAPOUR PRESSURE (Pa) 40°C
ethane	-183	-87	1.04		3.0	15.5	515	T1	IIA	30.07		
ethanethiol	-148	35	2.11	-20	2.8	18	295	T3	IIA	62.13		
ethanol; ethylalcohol	-144	78	1.59	12	3.3	19	425	T2	IIA	46.07	7965	17991
2-ethoxyethanol		135	3.1	95	1.8	15.7	235	T3	IIB	90.12		
ethoxyethyl acetate		156	4.6	47			380	T2	IIA	132.16		
ethyl acetate	-83	77	3.04	-4	2.1	11.5	460	T1	IIA	88.11	12618	25064
ethyl acetoacetate		180		(84)			295	T3	IIB			
ethyl benzene	-95	135	3.66	15	1.0	6.7	431	T2	IIA	106.17		2866
ethyl chloride	-136	12	2.22		3.6	15.4	510	T1	IIA	64.52		
ethyl cyclobutane			2.0	<-16	1.2	7.7	210	T3	IIA			
ethyl cyclohexane		131	3.87	14	0.9	6.6	262	T3	IIA	112.22		
ethyl cyclopentane	-80	103	3.4	1	1.1	6.7	260	T3	IIA	98.18		
ethyl formate		54	2.55	<-20	2.7	16.5	440	T2	IIA			
ethyl methyl ether		8	2.087		2.0	10.1	190	T4	IIB			
ethylene	-169	-104	0.97		2.7	34	425	T2	IIB			
ethylenediamine	8	116	2.07	34			385	T2	IIA			
ethylene oxide	-112	11	1.52		3.7	100	440	T2	IIB	44.05		
formaldehyde	-117	-19	1.03		7	73	424	T2	IIB	30.03		
formic acid		101	1.6	68			(520)	(T1)	IIA	46.03		
2-furaldehyde		161	3.3	60	2.1	19.3	315	T2	IIA	96.09		
heptane	-91	98	3.46	-4	1.1	6.7	215	T3	IIA	100.21	6092	12332
heptan-1-ol	-34	176	4.03	60					IIA	116.20		
hexane	-95	69	2.97	-21	1.2	7.4	233	T3	IIA	86.18		
hexan-2-one	-56		3.46	23	1.2	8	(530)	(T1)	IIA	100.16		
hydrogen cyanide		26	0.90	-18	5.6	40	(538)	(T1)	IIB	27.06		
hydrogen sulphide	-86	-60	1.19		4.3	45.5	270	T3	IIB	34.08		
hydrogen	-259	-253	0.07		4.0	75.6	560	T1	IIC	2.016		
isobutyl alcohol	-108	107	2.55		1.7	10.9	408	(T2)	IIA	74.12		
isopropyl nitrate		105		20	2.0	100	175	T4	IIB	105.09		
isopropyl alcohol	-86	83	2.07	12	2.0	12	425	T2	IIA	60.10	6019	14225
kerosine		150		38	0.7	5	210	T3	IIA			

FLAMMABLE MATERIAL	MELTING POINT °C	BOILING POINT °C	RELATIVE VAPOUR DENSITY	FLASH POINT °C	FLAMMABLE LIMITS % VOL		AUTO[3] IGNITION TEMP °C	T-CLASS[3]	APPARATUS GAS GROUP	MOLECULAR (FORMULA) WEIGHT	VAPOUR PRESSURE (Pa)	
					LEL	UEL					25°C	40°C
methane (firedamp)	-182	-161	0.55		5.0	15	595	T1	I	(5) 16.04		
methane (industrial)												
methanol	-98	65	1.11	11	6.7	36	455	T1	IIA	32.04	16671	35089
2-methoxyethanol	-86	124	2.63	39	2.5	14	285	T3	IIA			
methyl acetate	-99	57	2.56	-10	3.1	16	475	T1	IIB	74.08	28830	54081
methyl acetoacetate		170	4.0	67			280	T3	IIA	116.12		
methyl acetylene		-23	1.4		1.7				IIB			
methyl acrylate	<-75	80	3.0	-3	2.8	25			IIB	86.09		
methyl cyclohexane	-127	101	3.38	-4	1.15	6.7	260	T3	IIA	98.19		
methyl formate	-100	32	2.07	<-20	5	23	450	T1	IIA	60.05	83438	
2-methyl propan-1-ol	-108	107	2.55		1.7	10.9	408	(T2)	IIA	74.12		
methylamine	-92	-6	1.07		5	20.7	430	T2	IIA	31.06		
morpholine	-3	128	3.0	(40)			(310)	(T2)	IIA	87.12		
naphtha		35	2.5	-6	0.9	6	290	T3	IIA			
naphthalene	80	218	4.42	77	0.9	5.9	528	T1	IIA	128.17		
nitrobenzene	6	211	4.25	88	1.8		480	T1	IIB	123.11		
nitroethane	-90	115	2.58	27			410	T2	IIA	75.07		
nitromethane	-29	101	2.11	36			415	T2	IIB	61.04		
1-nitropropane	-108	131	3.06	49			420	T2	IIA			
nonane	-54	151	4.43	30	0.8	5.6	205	T3	IIB			1406
octane	-56	126	3.93	13	1.0	3.2	210	T3	IIA	114.23	1859	4139
paraformaldehyde		25		70			300	T2	IIB			
paraldehyde	12	124	4.56	17	1.3		235	T3	IIA	132.16		
pentane	-130	36	2.48	<-20	1.4	8.0	285	T3	IIA	72.15	70912	119534
pentane-2,4-dione	-23	140	3.5	34	1.7		340	T2	IIA			
pentanol	-78	138	3.04	34	1.2	10.5	300	T2	IIA	88.15		
pentylacetate	-78	147	4.48	25	1.0	7.1	375	T2	IIA	130.19		
petroleum				<-20				T3	IIA			
phenol	41	182	3.24	75			605	T1	IIA	94.11		
propane	-186	-42	1.56		2.0	9.5	470	T1	IIA	44.10		
propanethiol								T1	IIB	76.16		

FLAMMABLE MATERIAL	MELTING POINT °C	BOILING POINT °C	RELATIVE VAPOUR DENSITY	FLASH POINT °C	FLAMMABLE LIMITS % VOL		AUTO IGNITION TEMP °C [3]	T-CLASS [3]	APPARATUS GAS GROUP	MOLECULAR (FORMULA) WEIGHT	VAPOUR PRESSURE (Pa)	
					LEL	UEL					25°C	40°C
propan-1-ol	−126	97	2.07	15	2.15	13.5	405	T2	IIB	60.10	2779	6985
propan-2-ol	−86	83	2.07	12	2.0	12	425	T2	IIA	60.10	6019	14225
propene	−185	−48	1.5	< −20	2.0	11.7	(455)	(T1)	IIA	42.08		
propylamine	−101	32	2.04		2.0	10.4	(320)	(T2)	IIA	59.11	42115	77478
pyridine	−42	115	2.73	17	1.8	12	550	T1	IIA	79.10		
styrene	−31	145	3.6	30	1.1	8.0	490	T1	IIA	104.15		
tetrahydrofuran	−108	64	2.49	− 17	2.0	11.8	224	T3	IIB	72.11	21617	40215
tetrahydrofurfuryl alcohol		178	3.52	70	1.5	9.7	280	T3	IIB			
toluene	−95	111	3.18	6	1.2	7	535	T1	IIA	92.14	3792	7885
triethylamine	−115	89	3.5	0	1.2	8	(190)	(T4)	IIA	101.19		
trimethylamine	−117	3	2.04	(45)	2.0	11.6	410	T2	IIB	59.11		
1.3.5-trioxane	62	115	3.11	35	3.6	29	254	T3	IIA			
turpentine		149			0.8							
xylene	−25	144	3.66	30	1.0	6.7	464	T1	IIA	106.17		

Notes and References

1 Data sources:

BS 5345: Part 1: 1989. Code of practice for selection, installation and maintenance of electrical apparatus in potentially explosive atmospheres.

ICI/RoSPA publication No IS91. Electrical Installations in Flammable Atmospheres.

Lange's Handbook of Chemistry, 13th edition. McGraw Hill. ISBN 0-07-016192-5.

Perry's Chemical Engineers' Handbook, 6th edition. McGraw Hill. ISBN 0-07-049479-7.

Entries in the table for vapour pressure have been calculated from data in Lange's Handbook of Chemistry using tabulated values provided and substituting in Antoine's equation.

2 A blank in the table means that data was not available. Expert advice should be sought.

3 Temperature data of auto ignition temperature and T-Class given in brackets () should be regarded as provisional data only, since no confirmation of the data has yet been published.

4 Sublimation temperature is given for boiling point.

5 Formula weight for industrial methane is given for CH_4.

APPENDIX 9

Impact Test Apparatus

338

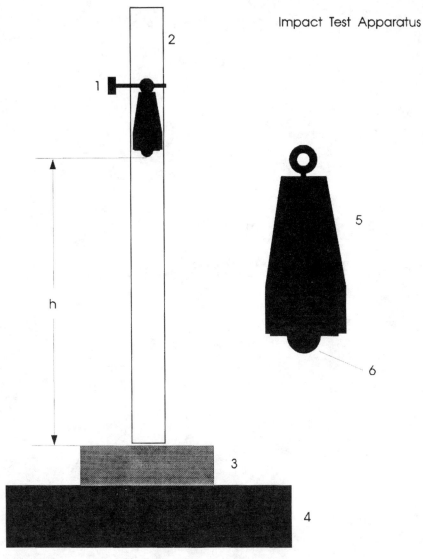

The steel mass of 1 kg (5) is secured in the guide tube (2) by the securing pin (1). On release of the securing pin, the mass falls to impact on the test piece (3).

The steel base (4) has mass of at least 20 kg. The impact head has a diameter of 25mm.

The impact head (6) is of polyamide for testing light transmitting parts, otherwise in hardened steel. Height h is adjusted to achieve required impact.

INDEX

Although the index includes a good cross reference to the main topics, not every mention of an indexed subject is included. For example, numerous references are made to Gas Groups throughout the book, but only the main explanatory references are noted in the index.

Maintenance   180, 197, 252-271
Marking (Certification)   99, 109, 118, 135, 152, 162, 163, 175, 197
Maximum surface temperature   11-13, 82, 286
Methane   8, 9
Methods of protection   46, 58, Section 2
Mineral insulating oil   112
Minimum ignition energy   11, 40
Minimum air gap   167
Mixtures of gases   36
Molecular weight   40, 330
Multicore cable   196, 223

N-Type protection   46, 92-101, 290
NEMA Code   212, 327
NFPA   14, 70, 310, 324
Non-hazardous area   6
Non-incendive   46, 93
North America   7

Obstructions to flame path   129-131
Oil immersion   52, 112
Oxygen enrichment   4, 5, 14

Periodic inspection & maintenance   263-267
PME. Potential Multiple Earthing   247-249
Portable apparatus   264, 270
Powder filling   53, 104-110
Pressure test   105, 115, 125, 144, 147
Pressure piling   133
Pressurization   49, 140-154
Propane   7, 8, 9
Protection height   108
Protective gas   140, 144, 146, 148, 152
PTB   206, 318
Purged apparatus   140-154

Quality assurance   65
Quartz   104

Radio-frequency ignition   5, 14
Records   268-269
Resistance, earth   261-262
Rotating machines   97, 134, 166, 169, 176, 177
Routine inspection & maintenance   263-266

Also available . . . .

Intrinsically Safe Instrumentation: a guide
by Robin Garside

This companion volume, now in its second reprint at Edition 2 covers the subject of intrinsic safety in detail, and provides a wealth of information on this specialist subject.

Available direct from Hexagon Technology Ltd.

ISBN 0 9508188 0 1

Training Videos:
Hazardous Areas: an introduction
The Methods of Protection
Intrinsic Safety

A valuable training tool for those new to the subject or requiring a time effective refresher.

Available direct from Hexagon Technology Ltd.

Hexagon Technology Limited provides specialist training and services to those industries involved with hazardous areas.